RELATIVITY
THEORY

RELATIVITY THEORY

Concepts and Basic Principles

AMOS HARPAZ

A K PETERS
WELLESLEY, MASSACHUSETTS

Editorial, Sales, and Customer Service Offices

A K Peters, Ltd.
289 Linden Street
Wellesley, MA 02181

Library of Congress Cataloging-in-Publication Data

Harpaz, Amos.
    Relativity theory : concepts and basic principles / Amos Harpaz.
      p. cm.
    Originally published: Boston : Jones and Bartlett, ©1992.
    Includes bibliographical references and index.
    ISBN 0-86720-220-3   ISBN 1-56881-026-1
    1. General relativity (Physics) I. Title.
QC173.6.H37   1993
530.1'1—dc20   93-4658
                                                         CIP

ISBN 1-56881-026-1

Scientific editor of the Hebrew version: *Dov Falik*

About the Cover

A radio photograph of a gravitationally-lensed object
... a photographic image of the object MG1131+0456, taken in radio wavelength ($\lambda = 2$ cm). The radio source MG1131+0456 is probably a quasar, containing a central compact core and two lobes located at opposite directions on either side, whose main luminosity is in the radio range of wavelengths. The bright (orange-yellow) spots are two images of the quasar's central core, created by the gravitational lensing of a gravitational field of a galaxy located between the quasar and us. The (green) ring is an image of one of the radio source lobes of the same quasar, located exactly behind the lensing galaxy. Such a ring is called "Einstein Ring," since Einstein predicted its appearance in 1936. The lenser (the lensing galaxy) is located on a line passing through the center of the ring. The main luminosity of the galaxy is in the optical range of wavelength. Hence it is not observed in radio range, in which this photograph was taken.

Cover photograph reprinted by permission from NATURE, 9 June 1988, Issue 6173, Vol. 333, cover. Copyright©1988 Macmillan Magazines Ltd.
Courtesy of Hewitt, J. N., et al., "Unusual radio source MG1131+0456: a possible Einstein ring."

Printed in the United States of America
96 95 94 93   10 9 8 7 6 5 4 3 2

Contents

Introduction   **1**

1. The Theory of Relativity   **5**
    References   12

2. Mathematical Introduction   **13**
    2.1. Vectors and Scalar Products   13
    2.2. Matrices   16
    2.3. Tensors   19

3. The Metric Tensor   **23**
    3.1. Coordinate Systems   23
    3.2. A Rectangular Cartesian System   26
    3.3. A Non-Rectangular Cartesian Coordinate System   28
    3.4. Non-Cartesian Systems   31
    3.5. A Spherical Polar Coordinate System   35
    3.6. Non-Cartesian Non-Rectangular Coordinate System   36

Conclusion   38
Comments   39

4. Space Dependent Metric   **41**
4.1. Curved Space   42
4.2. Apparent Curvature   44
4.3. Curvature of a Line   45
4.4. The Curvature of a Surface   47
4.5. The Properties of the Surface   49
4.6. The Components of the Metric Tensor   51
Summary   54
Comments and References   54

5. Four Dimensional Space   **57**
5.1. Four-Dimensional Coordinate System   57
5.2. The Interval   60
5.3. Space-Time Diagram   64
5.4. The Metric Tensor of the STR   66
5.5. Four-Dimensional Vectors   67
5.6. Lorentz Operator   69
5.7. The Rotating Disc System   71
5.8. General Gravitational Effects   73
Comments and References   75

6. The Principles of GTR   **77**
6.1. Action Over Distance   78
6.2. A Field   80
6.3. Momentum and Energy of a Field   82
6.4. Potential Energy   84
6.5. Inertial Mass and Gravitational Mass   85
6.6. The Principle of Equivalence   86
6.7. Bending of Light Trajectories   88
6.8. Shift of Wavelength of Light   91
6.9. Mach's Principle   95
6.10. Constraints on the Trajectories of Motion   100
Comments and References   106

7. Einstein's Equations   109
  7.1. A Geodesic Line   109
  7.2. Fermat Principle   113
  7.3. The Method of Variations   115
  7.4. Gravitational Potentials   120
  7.5. Coordinate Time   122
  7.6. Geodesics on a Rotating Disc   124
  7.7. Geodesic Triangle   128
  7.8. Einstein's Equations   129
  7.9. Non-Linearity of Einstein's Equations   132
        Comments and References   135

8. Schwarzschild's Solution   137
  8.1. The Exterior Schwarzschild Solution   139
  8.2. Application of Schwarzschild Solution to the Solar System   141
  8.3. Tests for the GTR   146
  8.4. Schwarzschild Radius   156
  8.5. Gravitational Collapse   157
  8.6. Effective Potential   159
  8.7. Black Hole   164
  8.8. Interior Schwarzschild's Solution   166
        Comments and References   166

9. Cosmological Solutions   169
  9.1. The Cosmological Constant   170
  9.2. Friedman's Model   172
  9.3. Co-Moving System   173
  9.4. The Metric Tensor of Friedman's Model   175
  9.5. The Constant k 178
  9.6. Length Measurement   180
  9.7. Hubble's Constant   181
  9.8. Time Evolution   183
  9.9. The Critical Density   186
  9.10. Red Shift of Light from Remote Galaxies   187
        Comments   190

10. Relativistic Astrophysics Phenomena   **191**
    10.1. Quasars   191
    10.2. High Density Matter   193
    10.3. Neutron Stars and Black Holes   194
    10.4. Gravitational Waves   197
    10.5. How Gravitational Waves are Created   200
        References   202

11. Epilogue   **203**

    Glossary   **207**

    Index   **223**

Introduction

This book intends to explain the principles of the General Theory of Relativity (GTR) for first year university students. It can also serve as a textbook for an introductory course on this subject.

Almost seventy years have elapsed since Einstein proposed his theory of relativity. The special theory, which deals with the specific case of non-accelerated (inertial) systems, has been already established as a part of the regular course of physics in high schools, but the ideas and the principles of the GTR still remain a "mystery" for the regular audience, even including advanced students of non-physics faculties.

It is my opinion that there is no objective reason for this situation. Its real cause is twofold: (1) the mystical image of this theory in the public eye, and (2) the lack of effort by the physics community to present the theory in simpler terms. From my experience as a teacher in high school and in a school of education, I have found that the problem in studying these kinds of theories consists of two components: first, the real difficulty of understanding complex situations which exist in nature and their mathematical description; second, the psychological difficulty in getting used to new ideas or new ways of thinking. Often have I seen intelligent students trying hard to find out what was it that they did not understand in the Special Theory of Relativity when they actually understood it quite well. It was really their subconscious refusal to accept its conclusions that made them so uneasy.

The ideas and the conclusions of the Theory of Relativity (both the special and the general) are indeed revolutionary, and this fact causes the psychological difficulty to dominate the problem. This also dictates that the way to change the situation is by supplying a wider population with a general understanding of the theory's ideas. This would create a suitable basis for including the ideas in the common world picture, and it would serve as a basis for advanced study of the topic for those who were interested.

Another difficulty concerns the mathematical tools needed to understand the theory. The GTR cannot be presented without using tensors. Upon encountering the concept "tensor" in a popular scientific text, lay readers usually close the book immediately, assuming that they have reached the limit of their ability to understand. Again, they do so without real justification. Our task here will be to remove the shroud of mystery from this concept and to show that although it is more complex, it is not different in principle from a vector, a concept already studied and used in high schools.

We shall begin with a mathematical introduction which, after reviewing vectors, deals with examples of scalar products of vectors, matrices as extensions of the vector concepts, simple operations between matrices and vectors, and finally arrives at the concept of tensors of rank 2.

The GTR will be presented here for the non-physicist reader. Emphasis will be on the transparency of explanations and illustrative examples, rather than on rigorous theorems and proofs. The professional reader may find repetitive explanations and generalizations without proof of intuitional understanding, but this is the price one must pay in order to prevent the book from being too heavy for the layman. The reward for this approach will be if the general principles and ideas of the GTR are indeed understood by a wider segment of the population.

A specific tensor, named "the Metric Tensor," plays a central role in the GTR. Its components are the variables of the theory, and these components, when determined, yield most of the information about the physical space. Hence, it is crucial for the understanding of the theory that this concept should be understood by the reader. Studying the contents of the theory and simultaneously getting acquainted with the new concept of the metric tensor is one of the main difficulties in learning the GTR.

To overcome this difficulty, two chapters are devoted to studying the role of the metric tensor using simple examples. Thus, the reader will gradually become acquainted with this mathematical object and the connection between its components and curvature of space before going into details of GTR.

Another important concept of the GTR is the "Geodesic line." We shall try to get some intuitional feeling for this concept and its connection with the properties of space. After the reader becomes familiar with these two important concepts, we can proceed into the consideration of the GTR itself. The principle of equivalence is studied as the basis for the connection between accelerated frames of reference and the geometry of curved spaces on the one side and the gravitational forces on the other side.

It is not expected that a non-physicist reader will be able to study solutions of partial differential equations (as Einstein's equations are). These equations will be presented in analogy with their Newtonian counterparts, concentrating on the differences between these equations. The contribution of the GTR will be understood by analyzing these differences and their influence on the resulting world picture. Such an approach will give the reader an overall picture of the GTR and of the way in which this theory presents the physical world. Having reached this point, we can enter into the subject of the most simple solutions of Einstein's equations (without actually solving the equations) and see how these solutions fit the information we have about the physical world. We shall also study some cosmological models resulting from solving Einstein's equations for the universe as a whole.

The general approach in this book is a literal one, and the use of mathematics has been reduced to a minimum (certainly too low in the opinion of some scientists), but still there are some points which the non-physicist reader will find hard to get through. Do not despair; the understanding of the rest of the text need not depend on the full understanding of each mathematical expression. I hope that the general picture will become clear to the reader who has the patience and who will make the effort to read this book to its conclusion.

This book may be used as a textbook for a course for the first degree in a university or college. Readers who are not interested in studying this topic professionally may read only the literal parts of the book and still acquire an understanding of the topic. Even a reader who intends to study the topic seriously will profit if he reads it once without going into the details of mathematics, and then repeats his reading with a rigorous study. In this way he or she might acquire a general view of the topic which might help in the detailed study.

The Hebrew version of this book was published in Israel in 1988. The present version was written during my stay at MIT, Cambridge, MA, in 1990, on a Sabattical leave from my Institution in Israel. I want to thank the people of the MIT for their hospitality during this stay.

CHAPTER ONE

The Theory of Relativity

Albert Einstein is known as the creator of the Theory of Relativity. Intuitively, one deduces that Einstein "invented" the concept of relativity and introduced its use in physics, or as the saying goes: "Since Einstein, we know that everything is relative."

Of course this is only an expression. The fact that the information we have on the physical world is relative to the basic situation of the observer was known for a long time. When we talk about relativity, we talk about the relative situation between the observer and the observed system, and it does not make any difference whether we mean the situation of the observer relative to the system or the situation of the system relative to the observer. The Newtonian mechanics, as it was formulated three hundred years ago, also included the mathematical formulation of the relativeness of phenomena, in the form called "Galilean relativity." If we try to define in brief what Einstein did to the principle of relativity, we shall say that in the Special Theory of Relativity (STR), Einstein solved contradictions which appeared in the application of the principle of relativity to the motion of light (and to the electro-magnetic effects in general). The solution of these contradictions led to far-reaching conclusions, which we shall deal with in detail later. This theory, proposed by Einstein in 1905, was called the "Special Theory" because it dealt with the specific case of motions in constant velocities only and did not deal with systems moving in acceleration relative to the observer. This means that it did not deal with measuring phenomena by an accelerating observer relative to the observed system.

The General Theory deals with the relativity of all kinds of motions and with measurements made by accelerated (or decelerated) observers. Here the conclusions were even more far reaching than those of STR: it was found that the treatment of the effects of gravitation can be included in the treatment of accelerated systems, and these effects can be interpreted as if they emerge from the situation of relative acceleration between an inertial observer and the observed system. Hence, the GTR, proposed fully by Einstein in 1915, includes also the theory of gravitation.

We shall use some simple examples to demonstrate the short description given above. Let us begin with motions having constant velocities, as they are described by the Galilean relativity. Since we deal with constant relative velocity between the observer and the observed system, the corrections to the description according to Einstein will be those emerging from the STR.

Consider a bus parked at a station. Let us designate the point where the rear edge of the bus is as point O. Two observers are supposed to report on the motions which take place in the system: observer A sits in the bus, and observer B stands parallel to him outside the bus. It is clear that as long as the bus is parked, the reports of both observers will be identical. Suppose now that the bus begins moving at time $t = 0$, with a constant velocity u. If the bus moves along a straight line, we can perform all our measurements along the line of the motion of the bus. Let us designate this line as the axis x. Until the time $t = 0$, the point marked by O was the same point for the two observers: it was the point where the rear edge of the bus was. Once the bus begins moving, observer A will assign it to the rear edge of the bus (which moves together with him) while observer B will assign it to the point on the ground where the rear edge of the bus was while the bus parked. To avoid confusion, let us mark the rear edge of the bus by O', and O will designate the point marked by observer B. The point O' will be the origin for the measurements of observer A, and all his measurements will be related to this point. (The same will be true for all the observers who stay with him in the moving system, the bus). The point O will be the origin for the measurements of observer B, and for the measurements of all the observers who stay with him in the rest system, the earth. From now on, we shall treat the earth and all the objects attached to it as the rest system, and the bus and all the objects staying in it as the moving system. All the entities determind by the observers staying in the moving system will be designated by a prime (′).

At the moment $t = 0$ both points O and O' coincide ($O = O'$). If we ask observer A to designate his position, he will report that according to his measurements he is located at some distance from point O'. Let us designate this distance by x'. On the other hand, when observer B marks the position of observer A, he will report the distance of observer A from the point O. Let us call this distance x. How do the distances x and x' related to each other? The distance x includes the distance x', and in addition it includes the distance of the rear edge of the bus from the starting point. This additional distance is the distance between O and O', and it is equal to the speed of the bus times the duration of the motion (the velocity is constant, and the motion began at $t = 0$), which is $u \cdot t$:

$$x = x' + ut. \tag{1.1}$$

The interrelation between x and x' is symmetric and hence:

$$x' = x - ut. \tag{1.2}$$

How will the two observers report on velocities? Suppose a ball is rolling in the bus with velocity v' (relative to the bus) and in the same direction of the bus motion. It is clear that observer A will report that the ball moves (relative to him) with velocity v'. The velocity of the ball as measured by observer B, however, consists of the sum of the velocity of the bus and the velocity of the ball relative to the bus:

$$v = v' + u \tag{1.3}$$

and of course:

$$v' = v - u. \tag{1.4}$$

Equations 1.1 to 1.4 are called the transformation equations of Galilei for the position and the velocity. Equations 1.3 and 1.4 are called "the law of addition of velocities." They can be accepted from equations 1.1 and 1.2 by differentiating them with respect to time, which means, by calculating the rate of change of the position on the condition that the time in the moving bus and on the earth is the same time. Stating that the time is an absolute entity (the time is the same in both systems and is independent of the measuring system) is actually a hidden assumption which lies at the basis of Newtonian mechanics. During the hundreds of years since Newtonian mechanics were formulated until the beginning of the twentieth century, this assumption was considered a self-evident one, and even today it is commonly accepted intuitively. Actually, one of the biggest difficulties in studying the STR is to accept the conclusion that time is not an absolute entity and that the results of time measurements depend upon the motion of the observer. Research on the evolution of the concept of time and its measuring was conducted by Professor G. Szamosi from the University of Windsor, Ontario, Canada, and published in his book "The Twin Dimensions."[1]

Until the end of the nineteenth century, there seemed to be no difficulty with Galilean transformation equations, and they suited the observations well. As was later discovered, the reason for this fact was that all the phenomena investigated were concerned with low velocities, except for the light motion. As for measurements concerned with light velocity, the degree of accuracy was so low that the contradictions between these equations and the observations were not observed. The problems arose when equations 1.3 and 1.4 were used in accurate experiments concerning the motion of light.

When one wants to relate these equations to light motion, one has first to determine what light is: is it a wave phenomenon or a corpuscular one? If the light is a corpuscular phenomenon, then its velocity (like the velocity of all other particles) depends upon the velocity of the light source. In such a case, by using the additional law for velocities one finds that the velocity of light relative to the observer equals the velocity of the light relative to the source, plus the velocity of the source relative to the observer.

If light is a wave phenomenon, then its additional law for velocities should be that of waves. When a wave moves in a medium, its velocity is defined relative to the medium and is determined by the properties of the medium. The wave velocity as measured by an observer is equal to the sum of the wave velocity relative to the medium and the velocity of the observer relative to the medium. At the beginning of the nineteenth century, it was established experimentally that light is a wave phenomenon, and hence people expected that the Galilean additional law of velocities for waves would be the correct law to use for light motion. The acceptance of the assumption that light is a wave phenomenon implied also the assumption that there is a medium in which the light moves as a wave. This medium was named "the Ether," and it was assumed that it fills the whole space and that it can be considered as an absolute rest system to which the motions of all objects can be related. Towards the end of the nineteenth century, scientists believed that light is a wave moving in the ether, and it was concluded that its motion could be treated according to the addition law of velocities for waves.

In 1887 the famous experiment of Michelson and Moreley was performed. In this experiment the scientists tried to measure the velocity of earth relative to the ether, where the technique of the experiment was based on the addition law of velocities for light. The degree of the precision of the experiment was very high, and significant results were expected. Yet the results of the experiment were null: no velocity of the earth relative to the ether was observed. Since then, the same experiment was repeated again and again with higher and higher precision, but always the same null results were obtained: the ether, to which the motion of earth was supposed to be related, was not found. The results of this experiment were considered a mystery; one that bothered Einstein greatly as he took his first steps in science.

The answer to the mystery was given by Einstein in 1905 in the form of the STR. This theory was based on two assumptions:

a.   There is no physical experiment by which a preference of one system over the other can be determined, if both of them move with constant velocity relative to each other.

b.   The velocity of light in *vacou* is constant and equal in all systems moving with constant velocities.

The meaning of the first assumption is that the principle of relativity is valid equally for all the systems moving with constant velocities, and there is no preferred system like the ether. The acceptance of the second assumption implies that the addition law for velocities should be corrected in such a way that the velocity of light will remain the same on transforming from one system to another. For this purpose the transformation equations for the position (1.1 and 1.2) were corrected too. From these corrections it followed that time could not be an absolute entity and that the time duration, measured for some given event, depends upon the situation of motion of the observer. (Actually, Einstein arrived first at the conclusion that the solution of the contradiction might be obtained only after abolishing the hidden assumption that time is an absolute entity. The correction of the equations was already done by him on the basis of the relativistic character of time.)[2]

The solution of the contradiction concerned with the relativistic character of light motion led to far-reaching conclusions, such as the contraction of moving bodies and time dilatation in moving systems. It was also found that the masses of moving bodies increase, and the conclusion that mass and energy are equivalent followed, as it is formed in the famous Einsteinian formula:

$$E = mc^2. \tag{1.5}$$

The important conclusions of the theory are presented by the equations which formulate them quantitatively:

$$L' = L_0\sqrt{1 - \frac{v^2}{c^2}}. \tag{1.6}$$

$$\Delta t' = \frac{\Delta t_0}{\sqrt{1 - \frac{v^2}{c^2}}}. \tag{1.7}$$

$$m' = \frac{m_0}{\sqrt{1 - \frac{v^2}{c^2}}}. \tag{1.8}$$

In all these formulae, the entities marked by $'$ are those connected with moving objects, while those marked by $_0$ are those measured when the objects are at rest.

The STR dealt with phenomena concerned with the behaviour of light relative to moving objects, and it actually included the solution for all the problems concerned with motions of electrodynamic entities. The name of the first article published by Einstein on this topic was "On the Electrodynamics of Moving Bodies." The topic of gravitational forces was not included in the STR, and Einstein knew that completing the theory demanded the inclusion of gravitational phenomena in it.

We have emphasized several times that the STR dealt only with systems moving with constant velocities, and thus did not include all kinds of motions. The general theory was supposed to generalize the principle of relativity over all kinds of motions, which includes accelerated systems. Einstein himself mentioned[2] that he was bothered by the fact that the STR defined a relation (equivalence) between mass and energy, but did not find any relation between inertial mass of objects and the gravitational influence on these objects. The solution became clear to him when he realized that gravitational phenomena, and phenomena connected with acceleration are equivalent, and it can be shown that part of the gravitational effects can be interpreted as effects emerging from acceleration. As a demonstration, we shall describe here the example of the elevator which was first presented by Einstein.

When subject to the action of a force, a free object will start accelerating and its velocity will increase as long as the force continues to act. Imagine an elevator staying in a gravitational field: if no force acts to stop it, it will fall in a free-fall motion in an acceleration emerging from the gravitational force. An observer staying inside the elevator and falling with it in a free-fall will not feel the gravitational force. If he drops some object, it will also fall in a free fall parallel to the observer, and thus will stay adjacent to the observer as if both of them stay in a gravitation-free region. This means that the fact that the observer is in a situation of free-fall (an acceleration) abolishes (for him) the effects connected with gravitation.

A counter example is the elevator staying in a force-free space (no gravitational force) and being pulled upward at a constant acceleration (its upward velocity increasing). An observer staying in this elevator feels as if a downward force acts on him, and he feels the pressure of the elevator floor on his feet, preventing him from sinking down through the floor: in other words, the fact that he stays in a situation of acceleration causes an effect similar to that caused by gravitation. If the observer does not look out (or if the elevator walls are opaque), the observer in the elevator will not be able to distinguish if the force he feels emerges from the fact that the elevator is accelerated or from the fact that he stays in a resting elevator in a gravitational field. We shall come back later to a detailed analysis of these phenomena, but it is already clear that at least for a part of the phenomena there is an equivalence between gravitational effects and effects connected with acceleration. In 1907 Einstein formulated this connection in a principle called the "Equivalence Principle." This principle suggests that from the physical point of view, there is an equivalence between effects emerging from the existence of a gravitational field and the effects emerging from the fact that the observer stays in an accelerated sytem.

This principle was the basis for the generalization of gravitational effects in the GTR. The general theory included both the principle of relativity for all kinds of motions and the gravitational effects. But the theory reached its final form only when Einstein found a geometric description of all the phenomena included in the GTR. In order to fully understand this last statement, we have to analyze the geometry of space and its mathematical description, and this we shall do in chapters 3, 4, and 5. For now, we shall say only that within the framework of the GTR, the phenomena connected with gravitation and those connected with acceleration appear as emerging from the geometry of space. Since gravitation emerges from the existence of matter in space, we shall expect also that the geometry of space will be determined by the distribution of matter in space.

The generally accepted picture is the following: The distribution of matter in space creates the geometry of this space, and this geometry determines the behaviour of objects, which seems to us as emerging from the forces of gravitation and acceleration. Chapters 6, 7, and 8 are devoted to the explanation of these last statements.

References

1.   Szamosi, G. *The Twin Dimensions*. McGraw Hill, New York, 1986.

2.   Einstein, A. "How I Created the Theory of Relativity." *Physics Today,* **35,** 8, 45.

Mathematical
Introduction

2.1.   Vectors and Scalar Products

A vector is a mathematical entity which describes quantities which have magnitude and direction, such as force, displacement, and velocity. A scalar is a number, and it describes quantities which have magnitude only, e.g., mass, temperature and energy.

A scalar product between two vectors is defined as a multiplicative operation between two vectorial quantities which yields a scalar quantity. The magnitude of the result is given by the product of the two vectors' lengths times the cosine of the angle between them.

As an example of a scalar product, let us calculate the work done by a force acting along a given path. The work done by the force is obtained by the scalar product of the force vector and the path vector. Let $\vec{F}$ and $\vec{R}$ be the force and the path vectors, respectively, and let W be the work. We solve this problem in a two-dimensional coordinate system. The equation for the work is:

$$W = F \cdot R \cos \theta \qquad (2.1)$$

13

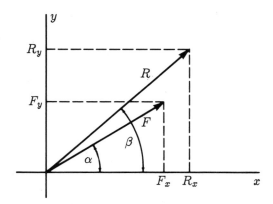

FIGURE 2.1.   The vectors $\vec{F}$, $\vec{R}$, in a two-dimensional Cartesian rectangular coordinate system. The angle between the two vectors is $\theta = \beta - \alpha$.

where θ is the angle between the directions of the force and the path. F, R, without the vector sign, represent the lengths of the vectors. In a vector notation, we write this equation:

$$W = \vec{F} \cdot \vec{R} \tag{2.2}$$

and this expression is equivalent to equation 2.1.

Let us calculate the product in a vector notation. We represent each vector by its components in a rectangular Cartesian coordinate system (Figure 2.1). The vector $\vec{F}$ is represented by its components F_x, F_y, and the vector $\vec{R}$ is represented by its components R_x, R_y:

$$\begin{aligned}
F_x &= F \cos \alpha; & F_y &= F \sin \alpha \\
R_x &= R \cos \beta; & R_y &= R \sin \beta
\end{aligned} \tag{2.3}$$

where α is the angle between $\vec{F}$ and the x axis, and β is the angle between $\vec{R}$ and the x axis.

We write the vector in a column form:

$$\begin{pmatrix} F_x \\ F_y \end{pmatrix}. \tag{2.4}$$

The scalar product between two vectors is performed by multiplying corresponding components in the two vector columns and summing the products:

$$\vec{F} \cdot \vec{R} = \begin{pmatrix} F_x \\ F_y \end{pmatrix} \cdot \begin{pmatrix} R_x \\ R_y \end{pmatrix} = F_x \cdot R_x + F_y \cdot R_y. \tag{2.5}$$

Indeed, substituting in equation 2.5 for the vectors' components from equation 2.3, we find:

$$\begin{aligned} \vec{F} \cdot \vec{R} &= F_x \cdot R_x + F_y \cdot R_y \\ &= F \cdot \cos\alpha \cdot R \cdot \cos\beta + F \cdot \sin\alpha \cdot R \cdot \sin\beta \\ &= F \cdot R(\cos\alpha \cdot \cos\beta + \sin\alpha \cdot \sin\beta) \\ &= F \cdot R\cos(\beta - \alpha) \end{aligned} \tag{2.6}$$

This expression equals equation 2.1 if we substitute: $\theta = \beta - \alpha$.

Another important example of a scalar product is the calculation of the length of a vector. The squared length of a vector is obtained by a scalar product of the vector by itself. (Let us call this product "a self-scalar product.") In order to calculate the length L of the vector $\vec{L}$, we write:

$$\begin{aligned} \vec{L} \cdot \vec{L} = \begin{pmatrix} L_x \\ L_y \end{pmatrix} \cdot \begin{pmatrix} L_x \\ L_y \end{pmatrix} &= (L \cdot \cos\alpha)^2 + (L \cdot \sin\alpha)^2 \\ &= L^2(\cos^2\alpha + \sin^2\alpha) = L^2 \end{aligned} \tag{2.7}$$

In the same way, the scalar products of three-dimensional vectors are performed. The calculations are more involved, but the results are essentially the same. A three-dimensional representation of the vector $\vec{F}$

is:

$$\vec{F} = \begin{pmatrix} F_x \\ F_y \\ F_z \end{pmatrix} \tag{2.8}$$

and the scalar product of two such vectors, $\vec{F}, \vec{R}$ is:

$$\vec{F} \cdot \vec{R} = \begin{pmatrix} F_x \\ F_y \\ F_z \end{pmatrix} \cdot \begin{pmatrix} R_x \\ R_y \\ R_z \end{pmatrix} = F_x \cdot R_x + F_y \cdot R_y + F_z \cdot R_z \tag{2.9}$$

This scheme of the scalar product may be performed in any number of dimensions.

2.2.   Matrices

A matrix is a mathematical entity which includes $n \times m$ components arranged in n rows and m columns (m components in each row). We shall deal with square matrices only, where we have $n \times n$ components—n rows by n columns. If we denote a matrix by **A**, its components are denoted by A_{ij}. In this notation $_{i,j}$ are indexes which assume the values 1 to n. A_{ij} is the jth component in the ith row.

An important axis in the matrix is the main diagonal from the top left to the bottom right. The components of this line are A_{ii}, where i runs from 1 to n. A symmetric matrix is one in which each component equals its mirror component with respect to the main diagonal. This is expressed by: $A_{ij} = A_{ji}$.

Let us define operations between matrices and vectors:

An equality between two matrices **A**, **B** exists if and only if each component in **A** is equal to the corresponding component in **B**:

$$A_{ij} = B_{ij} \tag{2.10}$$

The addition (subtraction) of two matrices **A**, **B** is performed by adding (subtracting) each component in **B** to (from) the corresponding component in **A**. The result is a third matrix **C**:

$$\mathbf{C} = \mathbf{A} \mp \mathbf{B}; \quad C_{ij} = A_{ij} \mp B_{ij} \tag{2.11}$$

A multiplication of a matrix **A** by a scalar f is performed by multiplying each component of the matrix by f, and the matrix **B** is obtained:

$$\mathbf{B} = f \cdot \mathbf{A}; \quad B_{ij} = f \cdot A_{ij} \tag{2.12}$$

Multiplication of a vector $\vec{B}$: $\vec{B} = \begin{pmatrix} B_x \\ B_y \\ B_z \end{pmatrix}$ by a matrix **A** yields a new vector $\vec{C}$. This operation can be performed only if the number of columns in the matrix equals the number of components in the vector. The operation is written:

$$\mathbf{A} \cdot \vec{B} = \begin{pmatrix} A_{11} & A_{12} & A_{13} \\ A_{21} & A_{22} & A_{23} \\ A_{31} & A_{32} & A_{33} \end{pmatrix} \cdot \begin{pmatrix} B_1 \\ B_2 \\ B_3 \end{pmatrix} = \sum_{j=1}^{n} A_{ij} B_j$$

$$\tag{2.13}$$

$$= \begin{pmatrix} A_{11}B_1 + A_{12}B_2 + A_{13}B_3 \\ A_{21}B_1 + A_{22}B_2 + A_{23}B_3 \\ A_{31}B_1 + A_{32}B_2 + A_{33}B_3 \end{pmatrix} = \begin{pmatrix} C_1 \\ C_2 \\ C_3 \end{pmatrix} = \vec{C}$$

Let us study this operation in some detail. The third term in equation 2.13, $\sum_{j=1}^{n} A_{ij} B_j$, indicates that for each row i we sum over the index j, which amounts to adding all the products of the matrix components in the ith row by the components of the vector, and this summation yields the ith component of $\vec{C}$, C_i. It means that C_1 equals the scalar product of the first row of **A** and $\vec{B}$. C_2 equals the scalar product of the second row of **A** and $\vec{B}$. C_3 equals the scalar product of the third row of **A** and $\vec{B}$. We conclude that for the multiplication of $\vec{B}$ by the matrix **A**, we consider **A** as a set of row vectors ($\vec{A}_i, i = 1, 2, 3$) and take the scalar products of these vectors and $\vec{B}$. This can be written:

$$C_i = \sum_{j=1}^{n} A_{ij} B_j$$

or:

$$\vec{C} = \begin{pmatrix} A_1 \cdot \vec{B} \\ A_2 \cdot \vec{B} \\ A_3 \cdot \vec{B} \end{pmatrix} \qquad (2.14)$$

where A_1 is the vector written in the first row of $\mathbf{A}$, etc.

A unit matrix $\mathbf{I}$ is one in which all the diagonal components are unity, and all the off-diagonal components are zero. In multiplying the vector $\vec{B}$ by the unit matrix, we find that multiplying the ith row of $\mathbf{I}$ with $\vec{B}$, only the ith component of $\vec{B}$ survives, so that the whole operation yields back the vector $\vec{B}$:

$$\mathbf{I} \cdot \vec{B} = \vec{B}. \qquad (2.15)$$

The product of two matrices $\mathbf{A}$, $\mathbf{B}$ yields a third matrix $\mathbf{C}$. This operation can be performed only if the number of columns of the first matrix equals the number of rows of the second one. (In square matrices, the number of the columns and rows of $\mathbf{A}$, $\mathbf{B}$, $\mathbf{C}$ are equal.) Let us write:

$$\mathbf{A} \cdot \mathbf{B} = \begin{pmatrix} A_{11} & A_{12} & A_{13} \\ A_{21} & A_{22} & A_{23} \\ A_{31} & A_{32} & A_{33} \end{pmatrix} \cdot \begin{pmatrix} B_{11} & B_{12} & B_{13} \\ B_{21} & B_{22} & B_{23} \\ B_{31} & B_{32} & B_{33} \end{pmatrix} = \sum_{J=1}^{n} A_{ij} B_{jk}$$

$$= \begin{pmatrix} A_{11}B_{11} + A_{12}B_{21} + A_{13}B_{31} & A_{11}B_{12} + A_{12}B_{22} + A_{13}B_{32} & \cdots \\ A_{21}B_{11} + A_{22}B_{21} + A_{23}B_{31} & \cdots & \cdots \\ \cdots & \cdots & \cdots \end{pmatrix}$$

$$= \begin{pmatrix} C_{11} & C_{12} & C_{13} \\ C_{21} & C_{22} & C_{23} \\ C_{31} & C_{32} & C_{33} \end{pmatrix} = \mathbf{C}$$

$$(2.16)$$

Studying this operation in detail, we find that we treat the first matrix **A** as a set of row vectors and the second matrix **B** as a set of column vectors. The scalar products of these vectors yield the components of **C**. Each component C_{ij} is formed by a scalar product of the row vector A_i (the i_{th} row of **A**) and column vector B_j. This can be written:

$$C_{ik} = \sum_{j=1}^{n} A_{ij} B_{jk} \qquad (2.17)$$

If **A** in this product is the unit matrix, then **C** = **B**; further, if **B** is the unit matrix, then **C** = **A**.

In matrix algebra we have an inverse matrix too—a matrix which on multiplying it by a matrix yields the unit matrix. Matrix $\overline{\mathbf{A}}$ is defined as the inverse matrix of the matrix **A**. $\overline{\mathbf{A}} \cdot \mathbf{A} = \mathbf{A} \cdot \overline{\mathbf{A}} = \mathbf{I}$. The entity $\overline{\mathbf{A}}$ exists only when the determinant of **A** is different from zero. (The concept 'determinant' is explained in the glossary.)

2.3. Tensors

"Tensor" is a mathematical concept, but we shall demonstrate first its use in physics. There are physical entities that cannot be described by scalars or vectors. A field of vectors might be sufficient to describe a force field, such as the electric field around a charged particle at rest. In this case we relate three numbers to each point in space. These three numbers are the components of a vector which describe the magnitude and the direction of the force acting on a charge at that point. Hence, relating a vector to each point in space provides all the information about the electric field in space.

Yet there are cases in which three numbers at each point are not sufficient to supply all the information needed to describe the physical state. An example of such a case is the stress field in a deformable body.

Consider a cubic differential element in such a body, which is located at the origin of a three-dimensional rectangular Cartesian coordinate system. Let us study the forces acting on the main three planes of this element, planes which are perpendicular to the coordinate axes. We find that the force acting on each plane can be decomposed into three components—one perpendicular to the plane and the two others parallel to it. We construct for each plane three entities which are the ratios of these components of the force and the area of the plane. The units of these entities are of force per unit area, which are the units of pressure. These three entities are the three components of the stress on the plane. The perpendicular component is called the normal component, and the two others are the shear components. (The normal component corresponds to the pressure in a physical system.) We have three stress components for each one of the three planes. All together we have nine components of the stress acting on the cube. Taking the limit of this configuration, when the volume of the cube tends to zero, all the nine components are defined at the same point. Hence, nine numbers are required to characterize the stress at each point in the body. We arrange these nine numbers in a matrix of 3×3, where each row in the matrix corresponds to one of the planes, and each column corresponds to one of the directions of the force acting on the plane. Denote the stress matrix by τ, then τ_{ij} represents the force acting on the ith plane in the jth direction. This mathemtical entity is a tensor. The stress components are displayed in Figure 2.2.

The physical entity represented above by the matrix is the stress tensor. The name "stress tensor" is a synonym, as the word *tensor* is derived from the root 'tense.' Probably, the first use of a tensor was to describe the stress in a physical body, and this might be the source of its name. Later, the tensor became an abstract mathematical concept, and the stress tensor is one of the ways this concept is used today. The stress tensor explained above is a tensor of rank 2. There might be various ranks for a tensor.

The formal definition for tensors given in many mathematical textbooks says: "A tensor is a mathematical form which transforms under special transformation rules which characterize tensors." This definition might not sound very meaningful at this point. It becomes more meaningful when the special transformation rules are given in detail, and thus the tensor is characterized more clearly. However, in later chapters, when we come to know a little more about transformation of vectors and get acquainted with some concrete examples of tensors, this definition will become more meaningful.

A tensor is characterized by its rank and the number of its components. The rank of a tensor is the number of the dimensions of the *form*. (Not to be confused with number of the dimensions of space.) The number

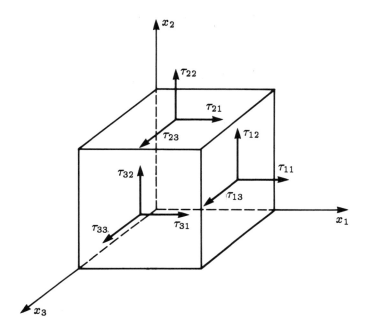

FIGURE 2.2.   Calculating the stress tensor τ. Three components of the force are acting on each surface of the cube; two components are parallel to it and one is perpendicular to the surface. τ_{ij} is the force components acting on the ith surface in the jth direction.

of components in each dimension of the form equals the number of the dimensions of the space on which this tensor is defined (like the number of dimensions of a vector). Hence the number of components of a tensor of rank m, in a space of n dimensions, is n^m. A square matrix of n dimensions is a form of two dimensions, and hence it describes a tensor of rank 2. The number of the components of this tensor is $n^2 = n \times n$. A vector of n dimensions is a form of one dimension, and hence it is a tensor of rank 1. The number of components of this tensor (the vector) is $n^1 = n$. A scalar is a form of zero dimension; hence it is a tensor of rank 0. The number of components of this tensor (the scalar) is $n^0 = 1$.

In this text we deal with tensors up to rank 2, which means that we deal with matrices, vectors and scalars.

The Metric Tensor

3.1. Coordinate Systems

A coordinate system is a mental device used to describe the location of objects in space. The name "coordinate" was given by Leibniz (G.W. Leibniz) who lived near the end of the 17th century. By locating the two end points of an object and calculating the distance between these locations, we can calculate the object's length. By locating the object at two time points (at the end points of a time interval), and dividing the distance between the two locations by the time interval, we can calculate the velocity of the object. Many other spatial and kinematic properties of objects (such as area, volume, acceleration) can be calculated by measuring locations in a coordinate system.

There are many types of coordinate systems—e.g., Cartesian, polar, spherical, elliptic—and each one of them is legitimate for the description of phenomena in nature. The choice between the different systems is arbitrary and is usually made on the basis of convenience. Since the coordinate system is a tool to describe natural phenomena, the nature of the phenomena cannot depend upon a specific choice of the coordinate system, nor can the physical results of the process described through the coordinate system method.

Since the choice of the system is arbitrary, we are able to transfer measurements and observations from one coordinate system to another. Such a transfer is called "transformation." The physical requirement of such transformations is that the geometric and physical properties of objects (described by [results of] observations) will be preserved through the transformations. Formulating this requirement mathematically: "The physical and geometrical properties of the physical system described will be invariant under the transformations."

Let us define more precisely the concept of geometrical and physical properties. The geometrical properties of an object are the quantities measured by measuring devices—lengths and angles between directions. (Entities of the kind of velocity are accepted by calculations made on measured quantities because velocity is not measured directly but is obtained by dividing distance by time.) The geometrical properties are "self properties" of the object and cannot vary with changes in the situation of the observer.

We use geometrical objects such as vectors to describe physical entities. The length of a vector describes (up to a coefficient of proportion) the magnitude of the physical entity represented by it, and the direction of the vector in space represents the direction of the action of that entity. For example, the velocity vector of an object represents the magnitude and direction of the velocity of that object. The requirement of invariance of geometrical properties means that the geometrical properties of the vector (its length and direction) will not change when we observe it from different points of observation (or from different coordinate systems).

A vector is given by its components along the axes of the coordinate system. We can rotate the coordinate system arbitrarily, and in the new system we shall have new values for the components of the vector along the newly positioned axes. The components of a vector are not geometrical properties of the vector. They depend upon the relation between the vector direction and the direction of the axes of the coordinate system. On the contrary, the vector length is a geometrical property, and we require that calculation of the vector length in the new system will yield the same value for its length as found in the original one. Further, an angle between two vectors is a geometrical property of their interrelation. Hence the evaluation of the angle in the new system must give the same result obtained from the observation in the original system.

The invariance of the geometrical properties of the vectors represents

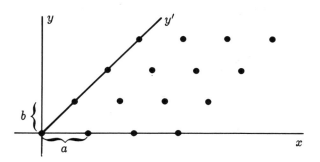

FIGURE 3.1.   Adjusting the angle between the axes of the coordinate system to the symmetry of the physical system. In the coordinate system x, y', the distances between the trees are the same along both axes.

the invariance of the physical properties of objects represented by those vectors. The invariance of physical properties through transformations from one coordinate system to another is the physical requirement of the transformations.

The most simple coordinate system is the "rectangular Cartesian system." (The name 'Cartesian' means that the axes of the system are straight lines. This name was given after R. Descartes, a French philosopher and scientist who was active at the beginning of the 17th century.) The axes in this system are straight lines and the angles between any pair of axes are right angles. In general, one can use any angle between pairs of axes, or non-straight lines as axes. Such a choice may cause a more complicated description, but sometimes such a choice may better suit the problem at hand.

Let us present an example which shows the convenience of choosing a special coordinate system. Suppose that the trees in an orchard are planted in lines so that the spacings between any two trees along the lines are constant (let us designate this spacing by a). The perpendicular distances between the lines are also constant, but different from a. Let us call this distance by b. It is found that the proportion between a and b is: $a = b \cdot \sqrt{2}$. Also each first tree in a line is planted at a distance b further than the first tree in the preceding line (see Figure 3.1). If we choose a rectangular Cartesian coordinate system to describe this orchard, the distances measured between neighbouring trees will be different along the different axes—they will be a along the x axis, and b along the y axis.

However, we can choose instead a Cartesian coordinate system, but not a rectangular one. If we have in the new system an axis y', (instead of y), whose inclination to the x axis is at an angle of $45°$, the distances between neighbouring trees which we measure along the y' axis will also be a, the same as the distances along the x axis. Many calculations of properties connected with the distances between the trees become simpler on using the system with the x, y' axes. By this choice we fitted the structure of the coordinate system to the geometrical structure of the orchard. On investigating crystallic materials, in which a very ordered structure exists with characteristic angles for each kind of a crystal, it is convenient to work with coordinate systems of the kind described above.

When describing surfaces, we can use a two-dimensional coordinate system (if the direction of the coordinate system fits the direction of the surface). For example, the motion of objects on the earth's surface can be described in a two-dimensional coordinate system, which is the system of latitudes and meridians. In this system the surface of the coordinate system (characterized by a constant radius of the sphere) coincides with the earth surface. In order to describe volumes, however, we have to use three-dimensional systems.

We shall try now to find a general way in which we can calculate a geometrical property of a vector, specifically its length. The (squared) length of a vector is obtained by a scalar product of the vector by itself. Such a method can be generalized later to find any physical quantity obtained by a scalar product of two vectors. By calculating the length of a vector in different coordinate systems, we shall study the use of the metric tensor, and we shall also find the metric tensors of different coordinate systems.

3.2.   A Rectangular Cartesian System

A rectangular Cartesian system is called a "flat system." In such a system the length L of a vector $\vec{k}$ is given by:

$$L^2 = a_1^2 + a_2^2 + a_3^2 \tag{3.1}$$

where a_1, a_2, a_3 are the components of the vector along the axes x, y, z correspondingly. If $\vec{k}$ connects the points P_1, P_2, then the components of $\vec{k}$ are given by:

$$a_1 = x_2 - x_1$$
$$a_2 = y_2 - y_1 \tag{3.2}$$
$$a_3 = z_2 - z_1$$

where x_1, y_1, z_1 are the coordinates of P_1 on the axes x, y, z, and the vector can be written:

$$\vec{k} = \begin{pmatrix} a_1 \\ a_2 \\ a_3 \end{pmatrix}. \tag{3.3}$$

The coordinate x_1 is the projection of P_1 parallel to the plane y, z on the x axis. The coordinate y_1 is the projection parallel to the plane x, z on the y axis, etc.

The expression (3.1) is a simple one (it can be interpreted as the Pythagorean formula in a three-dimensional space), and its form will not change if we perform a transformation to another coordinate system, rotated and/or translated relative to the original system, subject to the condition that the new system is a flat one too. Usually, the components of the vector change through such transformations, but the form of equation 3.1, where the squared length of the vector is given by a sum of the squares of its components, will remain the same. Moreover, the numerical value of the squared length will be preserved.

As we shall see later, if we describe the same vector in a non-rectangular coordinate system, then the expression for L^2 will also include mixed products of the components, multiplied by some coefficients of the angles between the axes. If we describe the same vector in a non-Cartesian system (like the geographic system of coordinates on the earth surface) or in a system in which we have different scales along the different coordinate axes, then in the form of L^2 there are coefficients different from unity with the squared terms of the form, and these coefficients could be functions of the locations.

Let us look for a mathematical tool which will generate the coefficients and the mixed terms whenever they are needed. This tool is called the "metric tensor." We shall give here some examples to demonstrate the role of the coefficients and the additional terms in the form of L^2 and to show how they are constructed with the help of the metric tensor.

3.3.   A Non-Rectangular Cartesian Coordinate System

Let us begin with a two-dimensional coordinate system in a plane. Our system consists of two straight line axes, U, V, and the angle between these two axes is some angle, α. We are looking for an expression for the length L of a vector $\vec{k}$ which connects the two points P_1, P_2, where the coordinates of these points are given along the axes U and V.

If α is a right angle, L^2 is given by:

$$L^2 = a_1^2 + a_2^2 \qquad (3.4)$$

where a_1, a_2 are the vector components along U, V, and are calculated in the way used to calculate expression 3.2.

We can observe that the expression 3.4 was obtained by a self scalar product (a scalar product of the vector by itself) of the vector $\vec{k}$. If the angle α is not a right angle, the expression for L^2 will be given by the cosine theorem (which is a generalization of the Pythagorean formula for a general triangle):

$$L^2 = a_1^2 + a_2^2 + 2 \cdot a_1 \cdot a_2 \cdot \cos \alpha. \qquad (3.5)$$

With the help of Figure 3.2, we can see how the components a_1, a_2 are calculated in such a system[1]: u_1 is the projection of P_1 parallel to the axis V on U, and v_1 is the projection of P_1 parallel to the axis U on V. In the same way, we calculate u_2, v_2 which are the projections of P_2 on U, V. We obtain the vector components a_1, a_2 in this system in the same way as in the previous case:

$$\begin{aligned} a_1 &= u_2 - u_1 \\ a_2 &= v_2 - v_1 \end{aligned} \qquad (3.6)$$

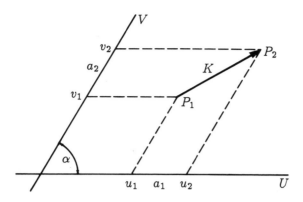

FIGURE 3.2.   Calculating the components of the vector $\vec{K}$ in a Cartesian system in the plane, where the angles between the two axes is α. The component along the U axis is the difference between projections of P_1, P_2, parallel to the V axis on U.

Again, for $\vec{k}$ we have:

$$\vec{k} = \begin{pmatrix} a_1 \\ a_2 \end{pmatrix}.$$

If we wish to interpret the expression 3.5 as a self scalar product of $\vec{k}$, we observe that this can be done if we multiply first one of the $\vec{k}$s by the matrix $\mathbf{A}$, where $\mathbf{A}$ is given by:

$$\mathbf{A} = \begin{pmatrix} 1 & \cos \alpha \\ \cos \alpha & 1 \end{pmatrix} \tag{3.7}$$

which amounts to writing:

$$L^2 = \mathbf{A} \cdot \vec{k} \cdot \vec{k} = (\mathbf{A} \cdot \vec{k}) \cdot \vec{k}. \tag{3.8}$$

The meaning of equation 3.8 is that we are multiplying the vector $\vec{k}$ (the right one in equation 3.8) by a vector obtained by the multiplication of the vector $\vec{k}$ by the matrix $\mathbf{A}$:

$$L^2 = \begin{pmatrix} a_1 + a_2 \cdot \cos \alpha \\ a_1 \cdot \cos \alpha + a_2 \end{pmatrix} \cdot \begin{pmatrix} a_1 \\ a_2 \end{pmatrix} = a_1^2 + a_2^2 + 2 \cdot a_1 \cdot a_2 \cdot \cos \alpha. \quad (3.9)$$

This form is a general one, and it also includes the special case in which α is a right angle. In this special case, $\cos \alpha = 0$, the matrix $\mathbf{A}$ is a unit matrix, and L^2 is equal to equation 3.4. This matrix $\mathbf{A}$, whose components are functions of the angles between the axes, represents the metric tensor of the space represented by this coordinate system.

The generalization of the form of L^2 to a three-dimensional non-rectangular coordinate system is simple in principle, but complicated for description because we cannot draw a three-dimensional figure. It can be interpreted as a continuation into a third dimension of the system described in Figure 3.2.

Let us denote the three axes (which are straight lines) by U, V, W, and the arbitrary angles α, β, γ as the angles between the pairs of axes U, V; V, W; W, U correspondingly. The three-dimensional vector $\vec{k}$ is the displacement vector between the points P_1, P_2, whose coordinates are u_1, v_1, w_1, and u_2, v_2, w_2. u_1 is the projection of P_1, parallel to the plane defined by V, W on U, v_1 is the projection of P_1 parallel to the plane defined by W, U on V, etc. The vector components are:

$$\begin{aligned} a_1 &= u_2 - u_1 \\ a_2 &= v_2 - v_1 \\ a_3 &= w_2 - w_1 \end{aligned} \quad (3.10)$$

$$\vec{k} = \begin{pmatrix} a_1 \\ a_2 \\ a_3 \end{pmatrix}$$

The squared length, L^2 is given by:

$$L^2 = a_1^2 + a_2^2 + a_3^2 + 2 \cdot a_1 \cdot a_2 \cdot \cos \alpha + 2 \cdot a_2 \cdot a_3 \cdot \cos \beta + 2 \cdot a_3 \cdot a_1 \cdot \cos \gamma. \quad (3.11)$$

Equation 3.11 can be obtained by a self scalar product of $\vec{k}$, if we multiply first one of the $\vec{k}$'s by the matrix $\mathbf{A}$, given by:

$$\mathbf{A} = \begin{pmatrix} 1 & \cos\alpha & \cos\gamma \\ \cos\alpha & 1 & \cos\beta \\ \cos\gamma & \cos\beta & 1 \end{pmatrix} \tag{3.12}$$

and the squared length is given by:

$$L^2 = (\mathbf{A} \cdot \vec{k}) \cdot \vec{k}. \tag{3.13}$$

The matrix $\mathbf{A}$ given in equation 3.12 represents the metric tensor of the general three-dimensional Cartesian space, and it characterizes the coordinate system chosen to describe this space. The components of the matrix are functions of the angles between the axes and are constants, and they do not change with location. It is easily observed that the form of $\mathbf{A}$ includes also the special case in which α, β or γ, or all of them are right angles.

Up to here we have dealt only with self scalar products because this very clear and transparent example shows the role of the metric tensor. In general, any scalar product of two vectors should be accompanied by the multiplication by the metric tensor of the space described by the coordinate system we are working with. For a scalar product between the vectors $\vec{b}$, $\vec{c}$, in a space in which the metric tensor is represented by the matrix $\mathbf{A}$, we write:

$$\vec{b} \cdot \vec{c} = (\mathbf{A} \cdot \vec{b}) \cdot \vec{c}. \tag{3.14}$$

In a rectangular Cartesian coordinate system, the metric tensor is represented by the unit matrix.

3.4. Non-Cartesian Systems

A non-Cartesian system is one in which not all the axes are straight lines. Let us first work with a two-dimensional system, a polar coordinate system on a plane.

In this system, one coordinate axis points in the radial direction from the origin, and the coordinates along this axis (assigned ρ) are the distances from the origin. The second coordinate is an angular one, measured along concentric circles around the origin. The zero line for the angular coordinate is a radial line chosen arbitrarily, and this coordinate, (assigned ϕ) measures the angle between the zero line and a radial line which passes through the point in question. The coordinate axes are orthogonal to each other everywhere, and hence there are no mixed terms in the form of L^2, and no off-diagonal components in the metric tensor.

The vector $\vec{k}$ is the displacement vector between two points P_1, P_2. The coordinate ρ_1 is the distance of P_1 from the origin, and ρ_2 is the distance of P_2 from the origin. The coordinate ϕ_1 measures the angle between the zero line and the radial line passing through P_1, and ϕ_2 measures the angle between the zero line and the radial line passing through P_2. One component of $\vec{k}$ is the term $\Delta\rho = \rho_2 - \rho_1$, and it is the radial distance between concentric circles passing through P_1, P_2. But the other component, $\Delta\phi = \phi_2 - \phi_1$, does not measure distance. Its units are not length units but pure numbers. The distance between two points, which have the angular coordinates ϕ_1, ϕ_2 and which lie on a concentric circle with radius ρ, is $\rho \cdot \delta\phi$, and it depends upon the radius of the circle.

The length of a curved line cannot be measured straightforwardly. For the process of measurement we divide the curved line into infinitesimal segments directed along the tangents to the line at each point. These segments should be small enough so that they can be considered straight lines. The total length of the curved line is obtained by calculating the length of each segment and summing the lengths of all the segments. This method will be used whenever we work with curved coordinate systems. In non-Cartesian systems we work with infinitesimal quantities, considered as differentials, and the summation of all the segments of which the line is composed is performed by integration. In Figure 3.3 we show how these coordinates are measured and how the vector components are calculated:

The displacement vector in curved space should be given in a differential form $\vec{dk}$, with the components $d\rho$, $d\phi$:

$$\vec{dk} = \begin{pmatrix} d\rho \\ d\phi \end{pmatrix} \tag{3.15}$$

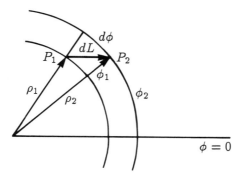

FIGURE 3.3.   Calculating the components of the vector $\vec{dL}$, in a polar coordinate system in the plane. $d\phi = \phi_2 - \phi_1$; $d\rho = \rho_2 - \rho_1$.

$$d\rho = \rho_2 - \rho_1$$
$$d\phi = \phi_2 - \phi_1$$

The square of the length of the differential, dL^2 is:

$$\vec{dk}^2 = d\rho^2 + \rho^2 d\phi^2. \tag{3.16}$$

The last expression is obtained by a self scalar product of $\vec{dk}$ if one of the $\vec{dk}$'s is multiplied first by the matrix $\mathbf{A}$:

$$\mathbf{A} = \begin{pmatrix} 1 & 0 \\ 0 & \rho^2 \end{pmatrix} \tag{3.17}$$

which yields:

$$dL^2 = (\mathbf{A} \cdot \vec{dk}) \cdot \vec{dk}.$$

$\mathbf{A}$ in equation 3.17 represents the metric tensor of the space represented by this system, and the significant property of this tensor is that its components are not constants (as was the case in the Cartesian systems) but are functions of the coordinates (ρ in the present case). This property characterizes systems in which not all the axes are straight lines.

Such a coordinate system is convenient for the description of systems in which the motions of most objects are in a plane, around some common center, such as the motion of the planets around the sun. The symmetry of such a system is a polar circular symmetry.

A three-dimensional system can be constructed on the basis of the plane polar coordinate system by adding a third coordinate, z, whose axis is a straight line, perpendicular to the plane defined by ρ, ϕ, and passing through the origin. This system is called a cylindrical coordinate system, and its metric tensor is represented by the matrix $\mathbf{A}$:

$$\mathbf{A} = \begin{pmatrix} 1 & 0 & 0 \\ 0 & \rho^2 & 0 \\ 0 & 0 & 1 \end{pmatrix}. \tag{3.18}$$

This system is convenient for the description of phenomena which involve axial symmetry—such as the magnetic field around a long conducting wire. The polar plane system we dealt with in the preceding section represents a space which is a subspace^2 of the space represented by the cylindrical system, for the case in which the coordinate z is constant (or equal to zero). Another subspace of the cylindrical system is obtained by taking ρ as equal to constant. The space obtained is a cylinder surface with ρ being the radius of the cylinder. The two coordinates of this space are ϕ and z, and the differential displacement vector $\vec{dk}$ is:

$$\vec{dk} = \begin{pmatrix} d\phi \\ dz \end{pmatrix}.$$

The metric tensor of this space is represented by the matrix $\mathbf{A}$:

$$\mathbf{A} = \begin{pmatrix} \rho^2 & 0 \\ 0 & 1 \end{pmatrix} \tag{3.19}$$

and it is a subtensor of 3.18.

It is interesting to note that although this space seems to be curved, the components of its metric tensor are constant on the surface. We shall come back to this point when we deal with curvature of surfaces.

3.5.   A Spherical Polar Coordinate System

An important non-Cartesian system is the space represented by the spherical polar system. In this system we have one length coordinate, r, whose direction is radial, and it measures distance from the origin. The other two coordinates are angular ones: θ marks the angles of the latitudes measured from the north pole of the system (down to the south pole), in the range $0 \leq \theta \leq \pi$, and ϕ marks the angle of the longitudes (the meridians) from a zero line chosen arbitrarily, and its range is $0 < \phi \leq 2\pi$. (Note that the latitudes on the earth's globe are measured from the equator, northward and southward, but this method is ambiguous since for each value of θ there are two latitudes—one north of the equator, and the other south to it.)

The coordinates of two points P_1, P_2 are r_1, θ_1, ϕ_1 and r_2, θ_2, ϕ_2 correspondingly. The displacement vector $\vec{dk}$ which connects these two points is:

$$\vec{dk} = \begin{pmatrix} dr \\ d\theta \\ d\phi \end{pmatrix}. \tag{3.20}$$

The coordinate θ is measured along great circles on the sphere and the length element along this coordinate is $r d\theta$. The coordinate ϕ is not measured along great circles on the sphere but along parallel circles whose radii are $r \sin \theta$, and hence the length element along this coordinate is $r \sin \theta d\phi$. The square of the differential length dL^2 of the vector $\vec{dk}$ is:

$$dL^2 = dr^2 + r^2 d\theta^2 + r^2 \sin^2 \theta d\phi^2 \tag{3.21}$$

and the matrix $\mathbf{A}$ of the metric tensor required to obtain 3.21 from 3.20 is:

$$\mathbf{A} = \begin{pmatrix} 1 & 0 & 0 \\ 0 & r^2 & 0 \\ 0 & 0 & r^2 \sin^2 \theta \end{pmatrix}. \tag{3.22}$$

Here also we do not have off-diagonal components in the metric tensor, nor mixed products in dL^2. The reason for this is that the axes of the three coordinates r, θ, ϕ are orthogonal to each other everywhere. The components of the metric tensor are functions of the coordinates as the system is not Cartesian. This coordinate system is convenient for the description of physical systems involving spherical symmetry, such as the electric field around a point charge at rest, or for the description of the balance of the pressure gradient and the gravitational force in a non-rotating star.

If we deal with the two-dimensional space of the surface of a sphere (which is a subspace of the spherical space), our coordinates will be θ, ϕ, and r will be a constant over the surface. The matrix **A** of the metric tensor of this subspace will be:

$$\mathbf{A} = \begin{pmatrix} r^2 & 0 \\ 0 & r^2 \sin^2 \theta \end{pmatrix}. \tag{3.23}$$

Although r is constant in this expression, the components of the metric tensor are not constants. This is connected with curvature properties of this space.

3.6.   Non-Cartesian Non-Rectangular Coordinate System

As an example of such a system, let us treat an elliptic coordinate system in a plane. We use a family of lines which are concentric ellipses with a constant eccentricity ϵ. The location of a point P in this system is done by first choosing the ellipse on which this point is located and then locating the point on the ellipse. The origin is the common center of all the ellipses. So, the two coordinates of the system are a and ϕ, where a is the length of the semi-major axis of the ellipse chosen, and ϕ is the angle between the semi-major axis and the line joining the point P to the origin (see Figure 3.4).

The lines $a =$ constant are ellipses, and the lines $\phi =$ constant are straight lines emerging from the origin. The coordinates of P_1, P_2 are a_1, ϕ_1, and a_2, ϕ_2 correspondingly. The differential vector connecting P_1 and P_2 is

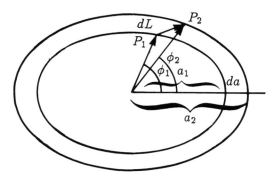

FIGURE 3.4.   Calculating the component of the vector $d\vec{L}$, in an elliptic coordinate system in a plane. $da = a_2 - a_1; d\phi = \phi_2 - \phi_1$.

$\vec{dk}$:

$$\vec{dk} = \begin{pmatrix} da \\ d\phi \end{pmatrix}.$$

In Figure 3.4 we show how the components of $\vec{dk}$ are calculated. The squared length of $\vec{dk}$ is:

$$dL^2 = \frac{1 - \epsilon^2}{(1 - \epsilon^2 \cos^2 \phi)} da^2 + a^2 \frac{(1 - \epsilon^2)[1 + \epsilon^2 \cos^2 \phi(\epsilon^2 - 2)]}{(1 - \epsilon^2 \cos^2 \phi)^3} d\phi^2$$

$$- \frac{2a\epsilon^2 \sin \phi \cos \phi(1 - \epsilon^2)}{(1 - \epsilon^2 \cos^2 \phi)^2} da \, d\phi \qquad (3.25)$$

The matrix $\mathbf{A}$ of the metric tensor is:

$$\mathbf{A} = \begin{pmatrix} \dfrac{1 - \epsilon^2}{1 - \epsilon^2 \cos^2 \phi} & \dfrac{-\epsilon^2 a \sin \phi \cos \phi(1 - \epsilon^2)}{(1 - \epsilon^2 \cos^2 \phi)^2} \\ \dfrac{-\epsilon^2 a \sin \phi \cos \phi(1 - \epsilon^2)}{(1 - \epsilon^2 \cos^2 \phi)^2} & \dfrac{a^2(1 - \epsilon^2)[1 + \epsilon^2 \cos^2 \phi(\epsilon^2 - 2)]}{(1 - \epsilon^2 \cos^2 \phi)^3} \end{pmatrix}$$

$$(3.26)$$

The components of the metric tensor are functions of the coordinates because the axes are not straight lines. There are off-diagonal components in the metric tensor and mixed products in the expression of dL^2 because the axes of the coordinates are not orthogonal to each other. We can easily see that the off-diagonal components vanish when $\phi = 0, \pi/2$, because along these lines the axes are orthogonal to each other. As the ellipses approach circles ($\epsilon^2 \to 0$), the form of the metric tensor approaches the one of the plane polar system, with $a \to \rho$. Usually such a system is not easy to work with, but it shows clearly the extreme case in which the system is both non-Cartesian and non-rectangular.

Another interesting case is the one in which we have different scaling along the different axes. Suppose that the length units along one axis are some basic unit times some constant k_1, and along the other axis they are the same unit times another constant k_2. In this case, the constants k_1, k_2 (or their ratio) will appear in the components of the metric tensor.

Conclusion

Although we have dealt here with self scalar products only, the same method is applicable in any scalar product of any two vectors. We perform a scalar product between two vectors by first multiplying one of them by the matrix of the metric tensor, and the vector obtained by this multiplication is then multiplied by the other vector.

Thus we have found one of the roles of the metric tensor, which is constructing the correct form of the scalar product of two vectors in a general coordinate system. We did not discuss the other roles of the metric tensor, nor did we show how this tensor is constructed or its components are calculated. We have already seen that the metric tensor is intrinsically connected to the properties of the space which the coordinate system represents, and actually the metric tensor characterizes the system.

In a flat system, the metric tensor is represented by the unit matrix. In a rectangular system, only diagonal components exist in the metric tensor. These components express the relations between the different axes, and if the coordinate system is non-Cartesian, the components will be functions of the coordinates. In a non-rectangular system there are off-diagonal components in the metric tensor and in general they are functions of the angles between the axes.

Comments

1. The vector components calculated in this way are called "Contravariant components." There are other ways to define the components of the same vector, like the "Covariant components," but we shall work only with the example given in the text.

2. Each n-dimensional space includes subspaces, whose number of dimensions is smaller than n. For example, a line is a one-dimensional space, and it is a subspace of a surface which is a two-dimensional space and is included in it. The surface is a subspace of a three-dimensional space and is included in it.

Space Dependent Metric

In the preceding chapter we dealt with the question of how we describe a given space through different coordinate systems. We have seen that we can arbitrarily choose any coordinate system to describe the space and the events occurring in that space. The choice is made according to considerations of convenience. Usually we try to choose a coordinate system which has the same type of symmetry as the problem at hand. For example, a gravitational field of a point mass has a spherical symmetry. All the points lying on the surface of a sphere whose center coincides with the point mass are equivalent in relation to the field. It is convenient to describe this case through a spherical coordinate system because it has the same symmetry as the described physical system. The system which describes the flow of a fluid in an infinite pipe has a cylindrical symmetry—all the points lying on a circle whose center is located on the pipe axis are equivalent when concerning the phenomena connected with the flow. It is convenient to describe such a case through a cylindrical coordinate system because it has the same symmetry as the described physical system.

We have also found that each type of coordinate system is characterized by a suitable metric tensor. However, it is clear that a choice of the coordinate system cannot change the physical properties of the space, nor can it influence the physical phenomena occurring in it.

In all the above, we have dealt with descriptions of nature through coordinate systems, assuming that space is independent of the events which take place in it. We have also assumed that space is always flat, homogenous and isotropic. Since these assumptions are subconscious, we have to explain what they mean.

A flat space is one that can be described by a rectangular Cartesian coordinate system whose number of dimensions is equal to the number of dimensions of the space, and in which the unit matrix represents the metric tensor (see Chapter 3). (Any curved space can be described by a rectangular Cartesian coordinate system if the number of dimensions of the coordinate system is larger than that of the space). A homogenous space is one in which all the points are equivalent, and there is no preference for an observer in choosing one location or another. An isotropic space is one in which all directions are equivalent, and there is no physical consideration preferring one direction over the other.

Usually, these assumptions are not explicit, but they always exist as hidden assumptions in the background of our thinking when we describe nature or when we formulate the physical laws. Under such assumptions the choice of the origin and the direction of the axes of the coordinate system is entirely arbitrary. The GTR, on the contrary, assumes the possibility that space is not flat, and that its properties could be different at different locations in space, and that these properties might change in time. Moreover, the GTR tries to relate the geometrical properties of the space to the field of gravitational forces which rule the space. Since changes in the geometrical properties of space must be reflected in suitable changes in the components of the metric tensor, we can study the evolution of the space properties by investigating the space and time dependence of the components of the metric tensor.

4.1.   Curved Space

In order to demonstrate the connection between the metric tensor and space properties, let us use the following method (for sake of convenience we begin with a two-dimensional space—a surface): Let us construct a network of points which will completely cover the entire space (the surface) we are going to investigate. At each point of the network we construct a local flat space. Even if the entire surface is not flat, each local flat space will be a plane, tangential to the surface at that point.[1] In each local flat space, the metric tensor will be represented by the unit matrix. The metric tensor represents the properties of the space to which it belongs. In a small neighbourhood of that point, the metric tensor of the local flat space represents also the properties of the real space to a good approximation.[2] Thus, the entire space is covered by a dense net of flat local spaces, whose (flat) coordinate systems suit the points at which they were constructed and their close neighbourhoods. Moving from one point to another, we may check how neighbouring flat systems fit each other, or how one flat system seems when observed from a neighbouring point.

If we proceed from one point to another along a "straight line" and align the axes of the local flat systems with the same angle relative to our direction of motion, we shall naturally expect that each local system will be a continuation of its predecessor. If the entire surface is indeed flat, each local system will seem flat when observed from neighbouring points, and all the local flat systems will unify to a one flat system of the entire space.

But there still exists the possibility (which is expected by the GTR) that the local flat systems will not fit to each other and that each system which was constructed as a flat one at a point will seem different from flatness when observed from neighbouring points.

Let us define two flat local systems S_1, S_2, constructed at the neighbouring points P_1, P_2 respectively. It might be that the system S_1, which was constructed as a flat system at the point P_1, will seem nonflat when observed from the point P_2. The same might be true for S_2 when observed from P_1. If S_1 seems from P_2 as deviating from flatness, these deviations could be changes of the angles between the major axes, or a change of scale along the different axes, or an inclination of S_1 with some angle relative to S_2. Observing continuously during a certain time, we might find that these changes could also evolve in time.

These differences between neighbouring systems could have an influence on information transferred from one system to another (from one point to another) and change the physical picture of the world when observing it from different locations in space. If we want to transfer results of measurements and observations from one local system to another, we cannot do it by a straightforward copying of the numbers. We have to construct proper equations for this transference, to ensure that the physical contents of the transferred information is not distorted through the transference. (Transferences of physical information from one system to another are called "transformations.") These equations reflect the differences between neighbouring systems, or, in other words, the deviations from flatness of S_1 when observed from P_2. The larger these deviations, the more unflat (curved) is our space.

If the space is unflat, there should exist at each point a *real* metric tensor, different from the tensor represented by the unit matrix. The real metric tensor characterizes the real space at that point. From the transformation equations which connect neighbouring local systems, we can calculate and construct the components of the metric tensor at that location, and the character of the equations will determine the character of the components. By this method we obtain at each point a local metric tensor, which represents the properties of space (curved or flat) at that point. If the network of points is infinitesimally dense, the differences between neighbouring local metric tensor will be small, and we can treat the components of the metric tensor as continuous variables of space. In this way a network of local flat coordinate systems join together to form a complete curved system which represents the entire space. The local original systems are flat, tangential to the total system at the points where they were constructed.

4.2.   Apparent Curvature

We have to distinguish between a real curvature and an apparent one. An apparent curvature is one which can be removed without changing the character of the space.

Let us look, for example, at a cylindrical surface. We now construct flat local systems at each point on this surface. When we move from one local system to another along a line which circles the cylinder, we shall find that each local system is inclined by a certain angle to its neighbour. Usually, such a property characterizes a curved space. We can, however, spread the cylindrical surface into a flat plane and this will not result in changes of the properties of geometrical objects which lie on this surface. This mean that the curvature had been removed, and the geometrical properties of the surface were not changed.

Such a curvature, which can be eliminated without changing the geometrical properties of the surface, is an apparent curvature. The same type of curvature exists on the surface of a cone: its surface can also be spread into a flat surface without changing the geometrical properties of the surface. On the other hand, a spherical surface cannot be spread in such a manner. We cannot spread it without changing its properties by stretching or cutting it into strips. (The attempts to sketch maps of the globe on flat maps, and the deformations created by doing it, are a good example of such a problem.) Such a curvature is a real curvature.

What is the criteria for distinguishing between a real curvature and an apparent one? On a cylindrical surface (or a conic one), we find that at each point there exists a direction along which the surface includes a straight line: on the cylinder, the line parallel to the cylinder axis; on the cone, the generating line. The local flat systems along this line join together to form a complete flat system. In contrast, on a spherical surface there is no direction in which a straight line can be included in the surface. We conclude that a curvature of a surface is apparent if at each point there exists at least one direction along which a straight line is included in the surface. If no straight line can be included in the surface at each point, the curvature is real.

Recall that while dealing with the metric tensor of a cylindrical surface we have found that on that surface the components of the metric tensor were constants, albeit different from unity $(1, \rho^2)$. In contrast, the components of the metric tensor on a spherical surface were not constants $(1, r^2, r^2 sin^2\theta)$. This is indeed a general rule: A space whose metric components are constants is a flat space or a space with an apparent curvature. By an appropriate transformation the metric tensor can be transformed into a unit matrix. The metric tensor of a space with a real curvature includes non-constant components and, as such, cannot be transformed into a unit matrix.

4.3. Curvature of a Line

In order to better understand the concept of curvature, let us study the curvature of a line (which is a one-dimensional space.) Let P_1, P_2 be two points close to each other on the line and ΔS be the arc (line segment) connecting these points. A slope of a line at a point is represented by the tangent to the line at that point. The tangence of the angle between the tangent and the x axis equals the derivative of the line's equation. If the line's equation is $y = y(x)$, then the slope of the line at a point P is $y'(x)$ (the derivative of y with respect to x at the point P), or:

$$y'(x) = \tan \alpha \tag{4.1}$$

where α is the angle between the tangent to the line at point P and the x axis.

The curvature q of the line is defined through the limit of the ratio between the angle difference $\alpha_2 - \alpha_1$, (at P_2, P_1) and the arc length ΔS, as ΔS tends to zero, i.e.,

$$q = \lim_{\Delta S \to 0} \frac{\Delta \alpha}{\Delta S} = \frac{d\alpha}{dS}. \tag{4.2}$$

The meaning of (4.2) is that the curvature of the line equals the derivative of the angle with respect to the arc length. This derivative can be constructed from the first and second derivatives of y with respect to x, and in standard geometry textbooks it is usually asserted[3] that:

$$q = \frac{d\alpha}{dS} = \frac{y''}{[1 + (y')^2]^{\frac{3}{2}}} \tag{4.3}$$

where y'' is the second derivative of y with respect to x.

Let us define R, the radius of curvature at a point. R is a radius of a circle that coincides with a line at the point (osculating circle), and it is equal to the reciprocal of the curvature:

$$R = 1/q = \frac{[1 + (y')^2]^{\frac{3}{2}}}{y''}. \tag{4.4}$$

On a straight line the second derivative vanishes. Hence the curvature of the straight line is nil, and the radius of its osculating circle is infinite. The stronger the curvature of the line is, the smaller the radius of its osculatory circle is.

4.4.   The Curvature of a Surface

Consider a surface—a two-dimensional space. The definition of the curvature of a surface at a point is related to the curvature of the lines passing through that point and included in the surface.

At a point P, where we want to calculate the curvature of the surface, let us construct the normal (a line orthogonal to the surface.) Then, construct a plane which includes the normal. This plane is called the "osculating plane," and it is orthogonal to the surface at the point P. The intersection of the osculating plane and the surface is a line. By writing the equation of this line, we can calculate its curvature (equation 4.3.) If we rotate the osculating plane around the normal as an axis, we get different intersections of the osculating plane and the surface. The curvature of each such intersection can be determined and calculated. Part of the lines might have positive curvatures, while other parts of the lines might have negative curvatures. (Sometimes all the lines might have positive curvatures, or all of them might have negative curvatures.) We shall find among all the lines one which will have the maximal curvature, and another which will have the minimal curvature. These two – the maximal and the minimal curvatures—are called the principal curvatures of the surface (at that point.) The product of the two principal curvatures is called the Gaussian curvature and is referred to as the curvature of the surface. Let us denote it by Q. The relation between the curvature of the surface and its radius of curvature is given by:

$$Q = 1/R^2; \qquad R = \sqrt{1/Q}. \tag{4.4.1}$$

As an example, let us examine the curvature of a spherical surface—a surface of a sphere.

On a sphere, the intersection of an osculating plane at a point with the sphere surface is a great circle which passes through the point. The curvature of a great circle equals $1/R$, (where R is the radius of the sphere.) As all the intersections of the osculating plane with the surface are equal, the two principal curvatures equal $1/R$, and the Gaussian curvature of the surface, Q, equals $1/R^2$. The radius of curvature of the surface equals the square root of the reciprocal of Q, and for the sphere it equals R (positive or negative.)

Recall that a surface with an apparent curvature is characterized by the fact that at each point on the surface there is at least one direction along which a straight line can be included in the surface. The curvature of a straight line is zero. Hence, at each point there is one (minimal or maximal) zero principal curvature. The product of the principal curvatures vanishes and the space is flat ($Q = 0$.) The radius of curvature of a surface equals $R = \sqrt{1/Q}$. Hence, in the case of apparent curvature it will tend to infinity.

What is the significance of the sign (positive or negative) of the radius of curvature? From equations 4.3, 4.4 it is clear that for a line the sign of R and q is determined by the sign of $y''(x)$. The second derivative of a line which curves upwards is positive. Hence its curvature is positive and so is its radius of curvature. Similarly, the signs of the curvature and of the radius of curvature of a line which curves downwards are negative. The signs of the curvatures of all the lines on a sphere are equal. In a coordinate system defined on the outer side of a sphere, all the lines on the sphere curve downwards, their curvatures are negative as are the signs of their radii of curvature. On the contrary, in a coordinate system defined at the inner side of a sphere, all the lines on the surface curve upwards, and their curvatures and their radii of curvatures are all positive. In both cases, the Gaussian curvature of the sphere Q, obtained through the product of two principal curvatures which have the same sign, will be positive. The radius of curvature of the sphere, obtained by taking the square root of the reciprocal of Q, will be real (positive or negative.)

There are cases in which the Gaussian curvature of a surface is negative, and the radius of curvature is imaginary. Consider the configuration of a saddle (Figure 4.1.) The lines passing through the saddle perpendicular to the line connecting the two hills of the saddle curve downwards, and their curvature is negative. The lines passing through the saddle connecting the two hills of the saddle curve upwards, and their curvature is positive. We find that one of the principal curvatures is positive while the other is negative.

The Gaussian curvature Q obtained through a product of the two principal curvatures will be negative, and the radius of curvature, which is the square root of the reciprocal of Q, will be imaginary.

There is no mystical significance to the fact that the radius of curvature is imaginary. The sign of the radius of curvature represents the direction of the radius relative to the surface. A line with positive curvature curves

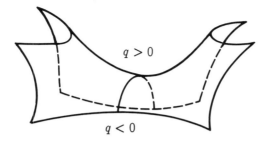

FIGURE 4.1. Calculating the Gaussian curvature of a saddle-like surface. One principal curvature is negative, while the other is positive. Gaussian curvature, obtained by the product of the two principal curvatures, is negative.

upwards, and its radius of curvature (which is the radius of the osculating circle which coincides with the line) is directed upwards. A positive sign for the radius of curvature means that the radius is directed upwards. The contrary is true for a negative sign for the radius of curvature. The sign of the radius of curvature of a surface at a point represents the signs of the radii of curvatures of all the lines included in the surface passing at that point. A positive sign for the radius of curvature of the surface at a point means that the radii of curvatures of all the lines passing through that point are positive. If some of the lines passing through the point have positive radii of curvatures, and the rest of the lines have negative radii of curvatures, then there is no general sign which can represent the signs of the radii of curvatures of all the lines passing through the point. In such a case the radius of curvature of the surface will be imaginary, which means that the general sign of the radii of curvatures at that point is undefined. The case of the saddle-like surface is an example of such a case.

4.5. The Properties of the Surface

We often used the statement "The properties of the surface." It is proper here to clarify what it means: A two-dimensional being, living in a two-dimensional space, is unable to *see* the nature of its space—the surface. (A curved surface is seen curved to an observer whose eyes are above the surface—which means that they are in a third dimension which is not included in the two-dimensional space and is actually unreachable for the two-dimensional being.) But the two-dimentional beings can identify the properties of geometrical objects in its space, and from these properties it can deduce the properties of the space.

Some of the geometrical properties are: the ratio of the circle circumference to its diameter, the ratio of the circle area to its diameter, the sum of the angles in a triangle, etc. It turns out that the sum of the angles in a geodesic triangle[4] (the concept "geodesic" will be explained in Chapter 7) is:

$$\alpha + \beta + \gamma = \pi + Q \cdot A \qquad (4.5)$$

where α, β, γ are the angles, and A is the triangle area. The value of Q and its sign determines whether the sum of the angles will be larger, smaller or equal to π. On a flat surface (a plane), Q vanishes and the sum of the angles of a triangle is exactly π.

The area of a geodesic circle is:

$$A = \pi r^2 - \frac{Q\pi r^4}{12} \qquad (4.6)$$

where A is the area of the circle, and r is its radius. On a plane $Q = 0$, and the area of the circle is πr^2. These formulae can be verified easily by using them on a spherical surface whose radius of curvature is the radius of the sphere R and its Gaussian curvature[5] is $Q = 1/R^2$.

Our two-dimensional being can measure the geometrical objects in its space, and by calculating the above mentioned relations it will be able to find the geometrical properties of its two-dimensional space without "seeing" it. Performing such measurements at an infinite number of points, it can find the curvature at each point and thus may draw a map of the curvature of the space. The map will show the curvature of the space at each point.

(By analogy, we can project conclusions for three-dimensional beings like ourselves, who live in a three-dimensional space. We cannot "see" deviations from flatness of our three-dimensional space, because we cannot exceed its bounds to a fourth dimension to perform our observations from there. Hence, even if our space is curved, we cannot observe it by direct "seeing." But from measurements of the geometrical properties of objects in this space, we may study the real properties of our space.)

The method we have suggested above for the two-dimensional being to map its space by measuring and calculating the curvature at each point in its space is equivalent to the method suggested in section 1 of this chapter. Recall that the meaning of that method was to calculate deviations from flatness at each point, and through these calculations map the space.

4.6.   The Components of the Metric Tensor

The components of the metric tensor are calculated from the deviations from flatness which exist at each point. Let us use the following notation: S is a local flat system tangential to the surface at point P. S' is the "real" system (defined in section 1) which coincides with the surface at the point P. As explained above, there are transformation equations which may be used to transform measurements and information from one system to another. The transformations relating quantities measured in S and S' reflect the deviations from flatness in the area around P. Let x_1, x_2 be the coordinates of the flat local system S (which is a rectangular Cartesian system), and y_1, y_2 be the coordinates of the real system S' (which will have in general a certain curvature.) The transformation equations relating the coordinates of S to those of S' will be expressions for x_1, x_2 as functions of y_1, y_2. The metric tensor of the flat system S is represented by the unit matrix. The metric tensor of the system S' is represented by the matrix $\mathbf{A}$. The components of $\mathbf{A}$ will be obtained from the transformation equations of the coordinates. They are constructed from the partial derivatives of the coordinates x_1, x_2 with respect to y_1, y_2. Let us construct them through the following procedure:

In the flat system S, the expression for the length of a vector dL is given by:

$$dL^2 = dx_1^2 + dx_2^2. \tag{4.7}$$

The quantity dL^2 was obtained by a scalar product of dL by itself (recall Chapter 2.) As dL^2 is a scalar, its value is invariant, and it is equal to its parallel quantity in S': $dL^2 = dL'^2$. Let us write:

$$dL'^2 = dx_1^2 + dx_2^2. \tag{4.8}$$

In order to express dL'^2 by y_1, y_2 we shall insert instead of x_1, x_2:

$$dx_1 = \frac{\partial x_1}{\partial y_1} dy_1 + \frac{\partial x_1}{\partial y_2} dy_2 \tag{4.9.1}$$

$$dx_2 = \frac{\partial x_2}{\partial y_1} dy_1 + \frac{\partial x_2}{\partial y_2} dy_2 \tag{4.9.2}$$

Taking the squares of (4.9.1, 2) and inserting in (4.8) we obtain:

$$dL'^2 = A_{11} dy_1^2 + A_{22} dy_2^2 + 2A_{12} dy_1 dy_2 \tag{4.10}$$

where we identified in (4.10) the components of the matrix $\mathbf{A}$ with the coefficients of the squares and products of the differentials in the following way:

$$A_{11} = \left(\frac{\partial x_1}{\partial y_1}\right)^2 + \left(\frac{\partial x_2}{\partial y_1}\right)^2; \qquad A_{22} = \left(\frac{\partial x_1}{\partial y_2}\right)^2 + \left(\frac{\partial x_2}{\partial y_2}\right)^2 \tag{4.11.1}$$

$$A_{12} = A_{21} = \frac{\partial x_1}{\partial y_1} \frac{\partial x_1}{\partial y_2} + \frac{\partial x_2}{\partial y_1} \frac{\partial x_2}{\partial y_2} \tag{4.11.2}$$

$\partial x_i / \partial y_j$ is the partial derivative of x_i with respect to y_j, and A_{ij} is the component of the matrix $\mathbf{A}$ at column j in row i. The components of the metric tensor of a real system in a space with higher number of dimensions are constructed through the same method.

A more compact form to write (4.11) is:

$$A_{ij} = \sum_{k=1}^{n} \frac{\partial x_k}{\partial y_i} \frac{\partial x_k}{\partial y_j} \tag{4.12}$$

where n is the number of the space dimensions, and k takes values from 1 to n. (The expresion 4.12 is suitable for calculation of the components of the metric tensor at any number of space dimensions.) An interesting exercise for the reader would be to derive the metric tensors given in Chapter 3 for the different coordinate systems, by using (4.11) or (4.12.)

From the expression (4.12) we can learn about the character of the components of the metric tensor. The components located outside the principal diagonal (A_{ij}, $i \neq j$) are obtained from mixed products of the derivatives with respect to different coordinates ($A_{ij} = (\partial x_k/\partial y_i)(\partial x_k/\partial y_j)$). This means that when the matrix representing the metric tensor is diagonal, there are no mixed products of different coordinates and no mixed products in the expression of the line element (Chapter 2.) When the mixed products do not vanish, it shows that there is some relation between the relevant coordinates in the picture which describes nature.

For example, when describing the events which take place in a rotating system of reference, the angular coordinate of each point is a function of time. Hence, we shall find in the expression for the square of the line element mixed product of differentials of the time and the angular coordinate, and we shall also find in the metric tensor an off-diagonal components which relate the two coordinates.

A coordinate which has no relation with any of the other coordinates is called a "Gaussian coordinate," and is defined as being orthogonal to all other coordinates. The concept "orthogonal" appearing here has a more general meaning than it has in the Euclidean geometry, and it is related also to systems including more than three dimensions. The fact that the components of the metric tensor include products of the partial derivatives of the coordinates has a bearing on the symmetry of this tensor.

As mentioned above, the off-diagonal components include mixed products of the derivatives with respect to different coordinates. The order in which the derivatives appear in the product makes no difference, and the order of the derivatives in the product might be interchanged without changing the value of the product. Hence $A_{ij} = A_{ji}$, or each off-diagonal component is equal to its symmetric component with respect to the principal diagonal, and the metric tensor is a symmetric tensor.

In a flat space, the transformation equations which relate S to S' are identities, and the metric tensor of the real system is represented by the unit matrix.

Summary

By the method described above we obtained a network of local real metric tensors, and we can map the entire space by the components of the metric tensors. This map will correspond to the map of curvature mentioned in section 1. At the points where the curvature vanishes in the first map, the metric tensor will appear as a unit matrix in the second map (or at most it will have constant components only.) In spaces where real curvature exists, the components of the metric tensor will be space-dependent variables.

We have found that the curvature of space and the components of the metric tensor are intrinsically related, and we conclude that the metric tensor characterizes uniquely the nature of space—whether it is curved or not, and to what extent is it curved. Later we shall find by examples how the curvature Q can be determined from the components of the metric tensor.

In most parts of this chapter we have dealt with a two-dimensional space, i.e. a surface. This was done for the sake of convenience and clarity. The descriptions of higher dimensional spaces are entirely analogous, but much more labourious and we leave it as an exercise for the interested reader. Detailed descriptions can be found in many textbooks of differential geometry such those of Wrede[6], Kreyszig[7], Do Carmo[4] and others.

Comments and References

1.   To construct a tangential plane at a point, we look at an infinite number of lines included in the surface, which intersect at the point. To each of these lines we construct a tangent. The assemble of all the tangents to the lines form a plane, and this is the tangential plane to the surface at the point.

2.   A "good approximation" means a description whose deviation from the real picture is small relative to the magnitude of the objects described. In the example at hand, the relevant region should be so small that the difference between the real space and the flat system will be very small relative to each phenomenon which we want to study.

3.   Coxeter, H. S. M., *Introduction to Geometry.* John Wiley and Sons Inc., New York, 1961.

4.   Do Carmo, M. P., *Differential Geometry of Curves and Surfaces.* Prentice-Hall, New Jersey, 1976.

5.   Equations (4.5) and (4.6) are correct for a surface whose curvature is constant. If the curvature is not constant, these equations should take their differential form and be integrated to give the correct answer.

6.   Wrede, R. C., *Vector and Tensor Analysis.* John Wiley and Sons Inc., New York, 1963.

7.   Kreyszig, E., *Introduction to Differential Geometry and Riemannian Geometry.* University of Toronto Press, Toronto, 1968.

CHAPTER FIVE

Four Dimensional Space

The description of the three-dimensional world gives us a static picture of the world, which is exact for the moment it was taken. If we want to follow the process of evolution we have to have a sequence of such static pictures, taken at a certain order in time, and the difference between successive pictures expresses the evolutionary process. Moving along such a set of successive pictures is like proceeding along the time axis, where each point along this "axis" represents a three-dimensional space at that given point in time.

5.1. Four-Dimensional Coordinate System

It seems natural to generalize time as a fourth coordinate, in addition to the three spatial coordinates, and thus to form a four-dimensional continuum which is a more complete description of the evolving universe. Usually we prefer to have the same units for all the coordinates. Choosing the length unit as the unit for all the coordinates, we have to multiply the time units by velocity. The velocity chosen for this purpose is the velocity of light in vacuum, c, which, as we already know from the STR, is an invariant

quantity in inertial systems. Hence the units along the time axis will be ct.
As the common length unit is 1 meter, then the appropriate time unit is the
time needed for light to proceed a length of 1 meter. A position vector R
in four-dimensional space-time[1] will be:

$$R = \begin{pmatrix} x \\ y \\ z \\ ct \end{pmatrix} = \begin{pmatrix} \vec{r} \\ ct \end{pmatrix} \tag{5.1}$$

where $\vec{r}$ is the three-dimensional position vector.

Proceeding along the time axis (while keeping the spatial coordinates
constant) means subsequent events in time, occurring at the same location.
In a four-dimensional coordinate system, an object at rest is moving on a
line parallel to the time axis.

As we cannot draw a four-dimensional figure, we shall forego two of
the spatial axes and draw a two-dimensional figure, where one axis will be
the time axis (ct), and the other will be one of the spatial axes (x). Thus we
can only describe one-dimensional motions. No generality is lost by this
choice if we describe only straight line motions because in this case we
can align the coordinate system in the direction of the motion, i.e., let the
motion coincide with the spatial axis drawn in the figure while the other
two coordinates (y, z) do not vary.

Figure 5.1 is a two-dimensional space-time figure for the coordinates
x, ct. Any object describes a line in this space-time diagram. This line
is called the "world line" of the object, and it represents the history of the
object in space.

In Figure 5.1, the line l_2 is the world line of an object which stays in rest
at the point $x = P$. The line l_1 represents the world line of an object moving
with a constant velocity in the x direction, which has passed through $x = 0$
at $t = 0$. The velocity of such an object (relative to the velocity of light) is
given by:

$$\frac{v}{c} = \frac{\Delta x}{\Delta(ct)} \tag{5.2}$$

where Δx is the distance (spatial interval) covered by the object during the
time interval Δt.

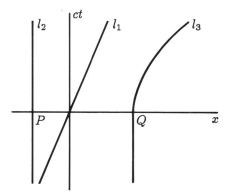

FIGURE 5.1.   World lines in a space-time diagram. l_2 is the world line of an object at rest at $x = P$. l_1 is the world line of an object moving with a constant velocity in the x direction. l_3 is the world line of an object which was at rest at $x = Q$, and at $t = 0$ began moving with a constant acceleration in the x direction.

As the object moves faster, the angle between its world line and the time axis will be larger. Each point in the four-dimensional space represents an event which occurs at a certain spatial point at a certain time point. If the spatial interval, Δx, between two points representing two events vanishes, it means that these two events occurred at the same location at different times. If the time interval Δct, between two events is zero, it means that they are simultanous events, occurring at different locations.

It comes out that the representation of the motions of objects in four-dimensional space is expressed in a form different from our intuitive perception. For example, let us follow an object which moves with a constant acceleration along the x axis. In a three-dimensional figure we shall see it moving along a straight line and the information about its changing velocity must be given as an additional information to the figure. This is because a three-dimensional picture is a momentary static picture, which does not show motions. Calculations of the motions can be done by comparing successive static pictures.

On the other hand, in a four-dimensional system, the trajectory of an accelerated object is represented by a line whose inclination to the time axis is increasing, as the angle between the object's world line and the time axis represents the velocity. The line l_3 in Figure 5.1 represents the world line of an object which was at rest at the location $x = Q$, and at the time $t = 0$ began moving with a constant acceleration in the x direction. The fact that no object can move faster than light implies that the maximum inclination available for a world line of a material object to the time axis is 45°, which is the inclination of a light world line.

The world line of an accelerated object will not be a straight line, and the changing inclination supplies the additional information about the changing velocity of the object. As a point in the four-dimensional space represents an event, so a line in this space represents a sequence of events, and the derivative of the line represents the velocity and its direction. When we deal with trajectories of objects in a gravitational field (geodetic lines), we shall come back to this topic.

5.2.   The Interval

We may try to construct scalar products of four-dimensional vectors and the metric tensor of this space. If we do it in the same way we used for the scalar product in the three dimensional space (Chapter 3, equation 3.7,9), we shall have for the "length" of the position vector R a sum of squares:

$$R^2 = x^2 + y^2 + z^2 + c^2 t^2. \tag{5.3}$$

We can ascribe a geometrical meaning (in the diagram) to this expression, but we do not know its physical meaning in the first place. We have to find physical considerations which will indicate to us what the form of the scalar product will be and what the metric tensor in this space will be. The appearance of the constant c (the velocity of light) in the sum of squares in equation 5.3 should not bother us. This velocity was chosen because, as postulated by Einstein, it is an invariant entity which does not vary on transforming from one inertial system to another. We shall use this property of c in our effort to define the scalar products in the four-dimensional space. The invariance of c may be formulated mathematically by the demand that for light motion:

$$\frac{\Delta x}{\Delta t} = \frac{\Delta x'}{\Delta t'} = \frac{\Delta x''}{\Delta t''} = c \tag{5.4}$$

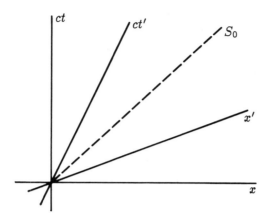

FIGURE 5.2.   A space–time diagram. A coordinate system O' is coupled to the object whose world line is l_1. The axis ct' coincides with l_1.

where x', t' are the coordinates of a system O', which moves with a certain velocity relative to the unprimed system O, and x'', t'' are the coordinates of the system O'', which, in turn moves with some other velocity relative to O.

This formulation represents the invariance of the velocity of light. In our search for an invariant quantity between the inertial systems which will have also a physical significance, we try to find an expression which will present the contents of equation 5.4. The principle of relativity asserts an equivalence of all the inertial systems, which means that any inertial observer can choose his own system of reference as the rest system for the description of the natural world. In Figure 5.2 we repeat Figure 5.1, but now we adjoin to the moving object whose world line is l_1 a system O', which moves with the same velocity relative to O. The time axis of O' (ct') coincides with l_1.

Thus, O and O' have the same origin. The dashed line S_o is the world line of a light ray passing through the origin with $(\Delta x/c\Delta t) = 1$. Relative to this trajectory x and ct are symmetric. The postulate on the velocity of light ensures that the same relation holds for O' too, namely $(\Delta x'/c\Delta t') = 1$. Let us use this demand to construct the axes of the system O'. The expression $(\Delta x'/c\Delta t') = 1$ asserts that the axes x', ct'

will also be symmetric relative to the world line of light (the same as x and t are symmetric relative to that line). The line ct' coincides with l_1, and the angle between x' and x is equal to the angle between ct' and ct. Thus we have found the axes of the system O' relative to the axes of O. They are inclined by a certain angle relative to the axes of O. We cannot obtain the axes of O' from those of O by a rotation of the system. The axis x' is inclined relative to x in the opposite direction of the inclination of ct' relative to ct. The transformation from O to O' is not a real rotation of the axes and is called pseudo-rotation.

Einstein has found that the transformation equations between O and O' are:

$$x = \gamma(x' + \frac{v}{c}ct') \tag{5.5.1}$$

$$y = y' \tag{5.5.2}$$

$$z = z' \tag{5.5.3}$$

$$ct = \gamma(\frac{v}{c}x' + ct') \tag{5.5.4}$$

where

$$\gamma = \frac{1}{\sqrt{1 - \frac{v^2}{c^2}}}.$$

We also have the inverse transformation equations:

$$x' = \gamma(x - \frac{v}{c}ct) \tag{5.5.5}$$

$$y' = y \tag{5.5.6}$$

$$z' = z \tag{5.5.7}$$

$$ct' = \gamma(-\frac{v}{c}x + ct). \tag{5.5.8}$$

These transformation equations (which are the transformation equations between inertial systems) are called Lorentz transformations, as they were evaluated by Lorentz before the STR was proposed by Einstein. The transformations of the coordinates whose axes are perpendicular to the direction of the relative motion between the two systems are identity transformations.

Let us use equation 5.4 to give a physical significance to an expression of the form of 5.3. It turns out that the form obtained for light motion will be:

$$c^2 dt^2 - (dx^2 + dy^2 + dz^2) = c^2 dt'^2 - (dx'^2 + dy'^2 + dz'^2) = 0 \quad (5.6)$$

(It should be mentioned that here dx^2 means $(dx)^2$ and not $d(x)^2$). The form 5.6 is called the "squared interval" and is usually referred to as $dS^2((dS)^2)$. The squared interval $dS^2 = c^2 dt^2 - dx^2 - dy^2 - dz^2$ is an invariant under Lorentz transformations. The term dt which appears in the interval is the time spacing between two events along the light ray trajectory, and the term dx, dy, dz appearing in the interval are the spatial spacings between the same two events. (Later we might use for brevity the form: $dx^2 + dy^2 + dz^2 = dl^2$). The value of the squared interval vanishes for light motion, but for two events which are not on a light ray trajectory, the interval might assume any value. From equations 5.5 we obtain the transformation equations for the differential dx, dy, dz, cdt and substitution of the transformed differentials into the expression of the interval shows that the squared interval is an invariant even when its value is different from zero.

A material object cannot move faster than light. Hence, for an interval between two events lying along a trajectory of a material object, the time interval, cdt, will always be bigger than the spatial spacing, and the squared interval is always positive:

$$dS^2 = c^2 dt^2 - (dx^2 + dy^2 + dz^2) > 0. \quad (5.7)$$

Such an interval is called "time-like interval." Two events connected by a time-like interval can communicate to each other (by information transmitted with the velocity of light or slower).

When the squared interval between two events is negative:

$$dS^2 = c^2 dt^2 - (dx^2 + dy^2 + dz^2) < 0 \qquad (5.8)$$

It means that the spatial spacing is bigger than the time interval, which means that the spatial distance between the two events is too large for a light ray to cover it during the time interval dt. The physical meaning of this statement is that no material influence, even a light energy, can connect these two events, and hence, no causal connection can exist between them. Such intervals are called "space-like intervals." As the squared interval is an invariant scalar, so its value is also an invariant, and its property (time-like or space-like) is an invariant too. The possibility for a causal connection between two events is an invariant, i.e., it is equal in all the inertial systems. Two events which cannot have a causal connection between them in one system cannot have such connection in any other system. (Some authors prefer to define the squared interval by: $dS^2 = dx^2 + dy^2 + dz^2 - c^2 dt^2$. Such a choice inverts the signs of the time-like and the space-like intervals.) In Figure 5.2 we adjoin to the object moving along the world line l_1 a coordinate system which moves with it, O'. This system is referred to as the "rest system" of the object, and all the entities related to the object and measured in this system are called "proper entities" of the object. For example, the length of the object measured in this system is its length when it is at rest, L_o, and is called its "proper length." A clock carried with this object is at rest in the system O', and it shows the time called the "proper time" of this object. What is the interval, dS along the object trajectory in the rest system of the object? As the object stays at rest in its own system, the spatial spacing in the interval vanishes, and the interval includes only the time spacing $dS^2 = c^2 dt^2$, i.e., the interval along the object trajectory in its rest system is proportional to the proper time along this trajectory. Usually the proper time is assigned by τ. Hence: $dS = cdt_o = cd\tau$.

5.3.   Space-Time Diagram

Figure 5.3 displays the different properties of space-like and time-like intervals: The world line of light rays define two cones in the figure, called

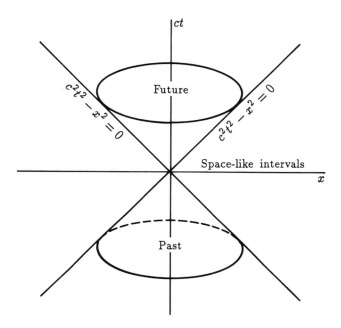

FIGURE 5.3.   Past and future cones in a space-diagram. The trajectory of an object which has passed through the origin must be included in these cones. Intervals which digress from these cones are space-like intervals.

"light cones" or "null cones" (as $dS^2 = 0$ along light ray trajectory). All the world lines of material objects passing through the origin have to be included inside these cones: The "future" of these objects is included in the upper cone, and the "past" of these objects is included in the lower cone. Intervals passing through the origin and digressing from these cones are space-like intervals, and they cannot represent world lines of any material object. The "past" cone includes all the events whose influence can reach the origin. The "future" cone includes all the events that can be influenced by an event located at the origin. Such past and future cones can be drawn for any event used as an origin, and thus show the possible connections between events in space-time.

5.4.   The Metric Tensor of the STR

The length of a vector (squared) was an invariant in a flat Euclidean space. This entity is a sum of the squares of the vector components, and hence it is positive by definition. The form of this invariant required that the metric tensor is represented by the unit matrix. The invariant of STR, dS^2, is not positive by definition, and it can assume positive as well as negative values. In order to obtain the (squared) interval given in equation 5.6 from a self-scalar product of the vector $dR = \begin{pmatrix} dx \\ dy \\ dz \\ cdt \end{pmatrix}$, we have to multiply it first by the matrix $\mathbf{A}$:

$$\mathbf{A} = \begin{pmatrix} -1 & 0 & 0 & 0 \\ 0 & -1 & 0 & 0 \\ 0 & 0 & -1 & 0 \\ 0 & 0 & 0 & 1 \end{pmatrix} \tag{5.9}$$

which represents the metric tensor of STR, i.e., $dS^2 = \mathbf{A} \cdot dR \cdot dR$. The components of this metric tensor are constants, although it is not represented by the unit matrix. Hence the space-time of the STR is a flat space. This space is called Minkowskian, after H. Minkowski, who investigated it on 1908. Sometimes it is called Pseudo-Euclidean space. If we want to use equation 4.12 to calculate the components of this metric tensor:

$$A_{ij} = \sum_{k=1}^{4} \frac{\partial x_k}{\partial y_i} \frac{\partial x_k}{\partial y_j} \tag{5.10}$$

we shall have to add a minus sign whenever $k = 1, 2$, or 3. Spaces over which equation 5.10 defines their metric tensor are called Riemannian spaces (after B. Riemann, 1854). The Euclidean space is a special case of Riemannian space.

Figure 5.4 shows[3] intervals in a space-time diagram.

The dashed line in this figure represents a trajectory of light for which the interval vanishes. The line S_1 is a hyperbola which is the locus of all the events for which their interval from the origin is given by:

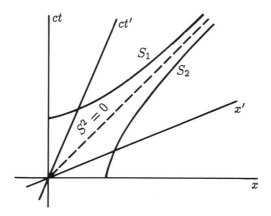

FIGURE 5.4. A space–time diagram (similar to Figure 5.2). The interval between the origin and any point on S_1 is: $S^2 = c^2t^2 - x^2 > 0$. The interval between the origin and any point on S_2 is: $S^2 = c^2t^2 - x^2 < 0$.

$S^2 = c^2t^2 - x^2 > 0$. The line S_2 is also a hyperbola which is the locus of all the events for which their interval from the origin is given by: $S^2 = c^2t^2 - x^2 < 0$. These values of S^2 are the same in the system O (with the coordinates x, ct) and in the system O' (with the coordinates x', ct'), and they are so in any other inertial coordinate system. In this diagram $|S_1^2| = |S_2^2|$.

5.5. Four-Dimensional Vectors

The important results of the STR are well known—the length contraction, time dilatation, and the equivalence of mass and energy. It comes out that all the calculations of the STR, including dynamics and electrodynamics, become simpler when using four-dimensional formalism, with the metric tensor represented by equation 5.9. For this purpose, four-dimensional vectors (four vectors) are defined, and they correspond to the three-dimensional ones. We already have introduced in equation 5.1 the position vector R. In parallel we define velocity four vector, which is called world velocity vector:

$$V = \gamma \begin{pmatrix} v_x \\ v_y \\ v_z \\ c \end{pmatrix} = \gamma \begin{pmatrix} \vec{v} \\ c \end{pmatrix} \tag{5.11}$$

where $\vec{v}$ is the three-dimensional velocity vector, c is the velocity of light, and γ is the Lorentz factor given in equation 5.5.

The momentum four vector is defined by:

$$P = \begin{pmatrix} \vec{p} \\ E/c \end{pmatrix} \tag{5.12}$$

where $\vec{p}$ is the three-dimensional momentum vector and E is the energy of the object whose momentum is $\vec{p}$.

The force four vector F is:

$$F = \gamma \begin{pmatrix} \vec{f} \\ \dfrac{dE}{cdt} \end{pmatrix} = \gamma \begin{pmatrix} \vec{f} \\ \dfrac{\vec{f}\vec{v}}{c} \end{pmatrix} \tag{5.13}$$

where $\vec{f}$ is the three-dimensional force and $(dE/dt) = \vec{f}\vec{v}$ is the power performed by this force.

Scalar products between vectors give scalars, which are invariant entities under transformations between inertial coordinate systems. We already know the invariant entity $dS^2 = (dR)^2 = (\mathbf{A} \cdot dR) \cdot dR$. We can obtain interesting results from scalar products between the four vectors given above. For example, the self scalar product of P, the momentum four vector is:

$$P^2 = \mathbf{A} \cdot P \cdot P = \frac{E^2}{c^2} - p^2.$$

According to the STR, this expression equals $m_o c^2$, where m_o is the rest mass of the object. It is an invariant and has the same value in all inertial systems.

A scalar product of the vectors P and R yields:

$$P \cdot R = \mathbf{A} \cdot P \cdot R = Et - \vec{p}\,\vec{r}.$$

The expression on the right wing is the argument of a wave function, and we find that this argument is also an invariant under Lorentz transformations.

5.6. Lorentz Operator

Let us use the transformation equations which we already know for the four vector R (given in equations 5.5) to find a general operator for the transformations of four vectors. Equations 5.5 can be written in the form:

$$R' = \begin{pmatrix} x' \\ y' \\ z' \\ ct' \end{pmatrix} = \begin{pmatrix} \gamma x - \gamma \frac{v}{c} ct \\ y \\ z \\ -\gamma \frac{v}{c} x + \gamma ct \end{pmatrix} = \Lambda \begin{pmatrix} x \\ y \\ z \\ ct \end{pmatrix} = \Lambda R. \qquad (5.14)$$

Equation 5.14 can be written in brief: $R' = \Lambda R$. The meaning of 5.14 is that the operator Λ (named Lorentz operator), in operating on the vector R, transforms it into R'. (R is the position four vector in the system O, and R' marks the same position in the system O'.) A straightforward calculation shows that Λ is a matrix of the form:

$$\Lambda = \begin{pmatrix} \gamma & 0 & 0 & -\gamma \frac{v}{c} \\ 0 & 1 & 0 & 0 \\ 0 & 0 & 1 & 0 \\ -\gamma \frac{v}{c} & 0 & 0 & \gamma \end{pmatrix} \qquad (5.15)$$

Operation of a matrix on a vector yields a new vector (equation 2.6). Hence, from equation 5.14 we find that a physical entity which was represented by a four vector in the unprimed system O is represented by a four vector (whose components are different) in the primed system O'.

Physical laws which were formulated in a four vector form will preserve
it and have the same form in any inertial system.

The form of natural laws in a form which is invariant under Lorentz transformations is called "covariant" form. The matrix Λ represents the Lorentz
transformation, and it operates in the same manner on all the quantities
represented by four vectors. The simple form of Λ, given in equation 5.15,
is related to the condition that the relative motion between the systems is
in the direction of the x axis. If we choose an arbitrary alignment for the
coordinate system where the relative motion between the systems is not
along the x axis direction, we have to write the transformation equations
in the form:

$$\vec{r}' = \vec{r} + \vec{v} \cdot \left(\frac{\vec{r} \cdot \vec{v}}{v^2}(\gamma - 1) - \gamma \frac{ct}{c} \right) \tag{5.15.1}$$

$$ct' = \gamma ct - \gamma \frac{\vec{r} \cdot \vec{v}}{c} \tag{5.15.2}$$

In these expressions $\vec{r}.\vec{v}$ means a scalar product between the three
vectors $\vec{r}$ and $\vec{v}$. The Lorentz operator in such a case is more complicated:

$$\Lambda = \begin{pmatrix} 1 + \frac{v_x^2}{v^2}(\gamma - 1) & \frac{v_x v_y}{v^2}(\gamma - 1) & \frac{v_x v_z}{v^2}(\gamma - 1) & -\gamma \frac{v_x}{c} \\ \frac{v_y v_x}{v^2}(\gamma - 1) & 1 + \frac{v_y^2}{v^2}(\gamma - 1) & \frac{v_y v_z}{v^2}(\gamma - 1) & -\gamma \frac{v_y}{c} \\ \frac{v_z v_x}{v^2}(\gamma - 1) & \frac{v_z v_y}{v^2}(\gamma - 1) & 1 + \frac{v_z^2}{v^2}(\gamma - 1) & -\gamma \frac{v_z}{c} \\ -\gamma \frac{v_x}{c} & -\gamma \frac{v_y}{c} & -\gamma \frac{v_z}{c} & \gamma \end{pmatrix} \tag{5.15.3}$$

By inspection of equations 5.15.1, 2, 3 we find that in the case for
which the relative velocity between the systems is along the x axis, these
expressions reduce to the simple form we have known before.

What are the transformation rules for tensors of higher rank?

It is found that tensor transformations are performed by operating on the
tensor with the same Lorentz operator, repeating this operation a number
of times which is equal to the rank of the tensor, namely, matrix, which is
a tensor of rank 2, should be operated twice by Λ. It can be proved that

this should be written:

$$\mathbf{B}' = \Lambda\, \mathbf{B}\, \Lambda$$

where $\mathbf{B}$ is the tensor to be transformed. The same rule also includes vectors and scalars. A scalar is a tensor of rank 0, it is an invariant, and it does not vary under Lorentz transformation. Vectors which are tensors of rank 1 are operated upon once by the Lorentz operator, and so on. The interested reader may find more details in textbooks dealing with this topic, e.g., Rindler[4].

5.7. The Rotating Disc System

Let us study in some detail an interesting case in four-dimensional space-time. A rotating system of reference is connected to a rigid disc which rotates at a constant angular velocity ω. Because of the cylindrical symmetry of the problem at hand, it is convenient to use a cylindrical coordinate system for the rotating disc, with the coordinates r, ϕ, z, ct (see equations 3.12). The coordinate r represents the projection on the disc plane of the distance from the origin to the point we are looking at. In the rest system, relative to which the disc rotates, the coordinates are the Cartesian ones: X, Y, Z, cT. The Galilean transformation equations between these two systems are:

$$\begin{aligned}
X &= r\cos(\phi + \omega t) \\
Y &= r\sin(\phi + \omega t) \\
Z &= z \\
cT &= ct
\end{aligned} \qquad (5.16)$$

The squared interval calculated by an observer in the rest sytem is:

$$dS^2 = -(dX^2 + dY^2 + dZ^2) + c^2 dT^2.$$

Substituting into this interval the differentials calculated from equations 5.16, we obtain the interval in terms of the differentials of the coordinates of the rotating system:

$$dS^2 = -(dr^2 + r^2 d\phi^2 + dz^2 + 2\omega r^2 d\phi dt) + (1 - \frac{r^2\omega^2}{c^2})c^2 dt^2 \quad (5.17).$$

The first three terms in this expression are familiar from the three-dimensional cylindrical system. The fourth term is a cross product of the differentials of the coordinates ϕ and t (it will result in an off-diagonal component in the metric tensor), and it expresses the coupling between the angular coordinate and time. This coupling is due to the angular motion of the system, and its meaning is that a point which is at rest on the disc (and asigned a fixed coordinate ϕ) is actually in a permanent angular motion relative to the rest system. What is the meaning of the coefficient of the last term $(1 - (r^2\omega^2/c^2))$? Recalling that the linear velocity of a point on the disc is $v = \omega r$, this coefficient turns out to be:

$$1 - \frac{r^2\omega^2}{c^2} = 1 - \frac{v^2}{c^2} = \frac{1}{\gamma^2} \quad\quad (5.18)$$

where γ is the Lorentz factor.

This reminds us of the transformation of the time coordinate in Lorentz transformation. Let us take the point of view of an observer who is at rest on the disc and rotates with it. For him $d\phi = dr = 0$ and the interval calculated in the disc system is:

$$dS^2 = (1 - \frac{r^2\omega^2}{c^2})c^2 dt^2 = (1 - \frac{v^2}{c^2})c^2 dt^2 \quad\quad (5.19).$$

An observer in the rest system outside the disc, who wants to compare his clock to a clock on the disc, will wait until the disc performs one complete rotation before he compares the clocks again, and hence, for him $dX = dY = dZ = 0$ and the interval in the rest system is $dS^2 = c^2 dT^2$.

Comparing the intervals of the two systems we find:

$$dt = \frac{dT}{\sqrt{1 - \frac{r^2\omega^2}{c^2}}} = \frac{dT}{\sqrt{1 - \frac{v^2}{c^2}}} = \gamma dT \quad\quad (5.20)$$

which is the Lorentz transformation for time. Although we did not introduce explicitly the Lorentz transformation, the Lorentz factor appeared as a result of the transformation we used. The observer on the disc will conclude that some effect in his system of reference (his rest system) causes his clock to slow, and that this dilatation depends upon its distance from the origin. Hence the coefficients in the expression of dS^2 (which are the components of the metric tensor of this system) are not constant, but depend upon location, and as we already know, this property characterizes curved spaces.

5.8. General Gravitational Effects

What other properties are connected to the fact that the disc system behaves like a curved space?

Suppose that there is a small object lying on the rotating disc. If the object is coupled to the disc (say, by a frictional force, as we are coupled to the earth's rotation), it will rotate with the same angular velocity as the disc. Due to the rotation the object will "feel" that it is subject to the action of the centrifugal force. The direction of this force is radial (outward along a radius of the disc) and its magnitude is:

$$F = mr\omega^2 = m\frac{v^2}{r} \tag{5.21}$$

where m is the object mass, and v is its linear velocity of rotation. An observer in the outside rest frame relates this force to the inertial property of the object which "tries" to continue its linear motion. But the observer on the disc, for whom the object is at rest, will look for a reason for this force. Sometimes these forces are called "fictitious forces" since they arise because of a special choice of the coordinate system. But for a being living on the disc (as we live in the rotating system of the earth), these forces are very realistic. We shall call them "inertial forces" as their source is the inertial properties of matter. The acceleration a, imparted to an object by this force, is:

$$a = \frac{mr\omega^2}{m} = r\omega^2. \tag{5.22}$$

We find that the acceleration does not depend on the object mass, so it will be the same for all material objects. We shall return to this point when we deal with the principle of equivalence.

Another important result is connected to the contraction of objects in the direction of motion. Rulers of standard unit length at rest on the rotating disc will contract relative to the length of their counterparts in the rest system if they are aligned along the direction of motion—the tangential direction. A rod aligned radially will not suffer any contraction. Now suppose that we measure the length of a circle on the disc (concentric with the axis of rotation) by counting the number of times the unit ruler enters in the complete circumference. The contracted rulers will be included more times in a complete circle than their counterparts in the rest system. Using the same method for the measurement of the radius whose direction is perpendicular to the direction of motion will give the same results in both systems. In the rest system the ratio of the circle circumference to its radius is exactly 2π. The observer in the disc system finds that this ratio is larger than $2\pi^5$. This property characterizes a curved space whose curvature is negative (equations 4.5, 6).

As was mentioned above, clocks on the rotating disc will be delayed by the Lorentz factor, which means extension of time units in the clocks on the disc relative to their counterparts in the rest system. The observer in the rest system will relate this time delay to the local velocity of the clock relative to his system. A rotating observer will not accept such an "excuse"—according to his observations the clock is at rest. He must connect the time delay he observes to the other curious phenomena he has already found: the centrifugal force and the ratio of the circle to its radius which is larger than 2π. His conclusion will be that in his system of reference there is some general effect which creates all the phenomena mentioned above: the appearance of inertial forces, the time dilatation, and the change of geometrical properties of geometrical objects. We have already mentioned that the cause for the change of geometrical properties is the curvature of space. The next move is to generalize the influence of that effect for the creation of all the other phenomena mentioned. Let us call all these phenomena "general gravitational effects" (in contrast to the regular gravitational effects which deal only with attraction between masses). The meaning of this name will become clearer later. We shall try to find out how the curvature of space influences physical phenomena and creates the general gravitational effects.

Comments and References

1. A vector notation with an arrow will designate the three-dimensional vectors.

2. An object thrown horizontally in a gravitational field describes a trajectory similar to that described in a four-dimensional diagram by an accelerated object moving with a constant acceleration in a straight line. The reason for this is that in such a motion we assume that the horizontal velocity of the object remains constant, hence the motion along the horizontal axis is proportional to time (where the coefficient of proportion is the initial velocity v_o). If we choose z as the horizontal coordinate and x as the vertical one, then, with an appropriate choice of the axes, we can write the equations of motions for x and z:

$$z = v_o t \qquad (5.23)$$

$$x = -\frac{1}{2}gt^2 \qquad (5.24).$$

Using equation 5.23 to substitute t into equation 5.24, we obtain an equation called the "equation of the trajectory" in which x is given as function of z. This equation does not involve time and it describes only the object's trajectory. We obtain:

$$x = -\frac{1}{2}g\frac{z^2}{v_o^2} \qquad (5.25)$$

which is an equation of a parabola. The parabolic character of the trajectory always appear when we deal with constant acceleration.

3. Wrede, R. C., *Vector and Tensor Analysis.* John Wiley and Sons, New York, 1963.

4. Rindler, W., *Special Relativity.* Oliver and Boyd, Edinburgh and London, 1966.

5. Moller, C., *Theory of Relativity.* Oxford University Press, London, 1966.

The Principles of GTR

The STR dealt with the equivalence of all the inertial frames of reference and constructed a covariant formulation of the physical laws which are invariant under transformations between these systems. But limiting the principle of relativity to inertial systems only (systems moving with constant velocities) is artificial. All kinds of motion are relative phenomena and, in principle, there is no reason to prefer inertial frames of reference over accelerated ones. A natural generalization of the theory is to impose the principle of relativity on all kinds of motions. This was done by Einstein in the GTR, where the basic assumption is that all frames of reference, including accelerated ones, are equivalent for the description of the natural world.

The GTR succeeded in achieving this goal by unifying several classes of phenomena, whose connections to each other are not immediately apparent. These are the gravitational phenomena, accelerated frames of reference, and the geometrical description of curved spaces. By gradually constructing the connections between these classes of phenomena, we shall be able to understand how the general design was attained.

6.1.   Action Over Distance

The gravitational interaction is given by the Newtonian law of the gravitational force:

$$\vec{f} = -G\frac{m_1 m_2}{r^2}\hat{r} \tag{6.1}$$

where $\vec{f}$ is the force (a vector), m_1, m_2 are the interacting masses, r is the distance between their centers, and G is the gravitational constant whose value (in MKS units) is:

$$G = 6.671 \times 10^{-11}\frac{\text{meter}^2 \times \text{newton}}{\text{kg}^2}.$$

In equation 6.1, $\hat{r}$ is a unit vector whose direction is the line connecting the two masses centers, and it points in the direction of the force. The minus sign denotes that the force is an attractive one. It is important to understand that the force acts equally on both masses, in the opposite directions: m_2 is attracted towards m_1 by the force $\vec{f}$, and m_1 is attracted towards m_2 by exactly the same force.

According to the law of action and reaction, this can be formulated as: m_1 attracts m_2, and the reaction force causes it to be attracted by the same force towards m_2. But both masses are symmetric with the respect to this force, and the same formulation holds with the interchange of the roles of m_1 and m_2. Equation 6.1 describes the situation exactly, as long as the system is static. But a dynamic situation, in which the masses are in motion relative to each other, may give rise to the question of how fast the interaction between the masses propagates.

To clarify the question, suppose a situation in which only m_1 is present in some region. At a certain moment we locate the second mass, m_2, at a distance r from m_1. The question is: will both masses feel the presence of each other simultaneously? Does the action of m_1 reach m_2 immediately (moving with infinite speed), or does it reach m_2 a little later (moving with a finite speed)? When does m_2 activate its action on m_1? If its action propagates with a finite speed, will m_1 feel the reaction to its action on m_2 immediately, or will the reaction be delayed by the time needed for a return trip over the distance between the masses? If so, what happens to the third law of Newton—the law of action and reaction? (The formulation of the third law is given in the Glossary.) The problem with the third law is that it deals with action and reaction between objects where the distance between them is finite. The law states that objects acts and feel the reaction to their action simultaneously.

The answer of Newton to the question presented above is very simple: He assumed immediate interaction over distance between objects with infinite speed. This results in simultaneity of interaction on both objects, and thus the validity of the third law is guaranteed.

To better understand the above statements, let us analyze what 'an interaction' is (an action between objects). When an object a exerts a force on object b, object b will respond. If it is free to move, it will acquire acceleration according to the equation of motion (the second law of Newton), and it will acquire momentum and kinetic energy as a result of the interaction. Thus, the interaction is characterized by the transference of momentum from the acting object (a) to the activated one (b). If object b is not free to move, a force develops which acts against the force performed by a, which prevents the transference of the momentum to b. (The two possibilities described above are the two extreme ones. Usually the situation will be an intermediate one, in which part of the acting force will be braked, and another part will act and impart momentum to object b). In both cases object a feels a reaction, which is the action of a force equal by magnitude to the acting force but opposite in direction. In the case that object b is free to move, the source of the reacting force is the inertial force of object b. In the case that b is not free to move, the source of the reacting force is the external force which prevents b from moving. In this description we ignored the time needed for the acting force and the reacting one to cover the distance between the objects (Newton's assumption about infinite speed for action and reaction).

The principles of the Theory of Relativity contradict Newton's assumption and its consequences. According to the Theory of Relativity, no material influence can move faster than light. When time lost its absolute character and was found to be a relativistic entity, the simultaneity also lost its absolute character. Events observed to be simultaneous in one frame of reference may seem to be occurring at different times in another system. The third law of Newton is valid only in interactions which take place at the same location and at the same time. Newton's assumption of immediate interaction over distance cannot hold together with the Theory of Relativity.

6.2.   A Field

The concept of the field replaces Newton's approach.

First, we assume that the gravitational interaction travels in space with a finite speed (the speed of light in vaccou). Second, the influence of the mass m_1 does not wait for the appearance of m_2 to start expanding in space. Using the preceding example in which only m_1 is located in a certain region, the space around it acquires the property that, when a second mass (m_2) is introduced anywhere into this vicinity, a force (according to equation 6.1) will act on it immediately. The potential to exert this force is related to the space (hence there is no need for the action to propagate from the location of m_1), and this property of space is induced by the presence of m_1. The potential of space to exert a force on any mass located in it is called the "gravitational field of force," and m_1 plays the role of the source of the field. It should be understood that the field was constructed by a time-dependent process, by propagation with a finite speed from the source, when m_1 was first brought to its location. Once the field is constructed, it will exist as long as the source is present. When the source is removed, the field will disappear, again with the same speed it was constructed.

A compact definition of the field is: "The field is the property of space. The magnitude of the field at a point equals the force which will act on a unit mass when located at that point." The units of the field are force per unit mass. Hence the force which will act on m_2 when introduced into the field equals the magnitude of the field times the mass m_2.

Denoting the field by $\vec{g}$ we find:

$$\vec{f} = m_2 \cdot \vec{g}. \tag{6.2}$$

Comparing equation 6.2 with equation 6.1, we find that the field $\vec{g}_1$ generated by m_1 as a source is:

$$\vec{g}_1 = -G\frac{m_1}{r^2}\hat{r} \tag{6.3}$$

where r is the distance from the source to the field point.

Equation 6.2 yields an interesting consequence. If the force acting on m_2 equals $m_2 \cdot \vec{g}$, then the acceleration imparted by the action of this force, according to the equation of motion is:

$$\vec{a} = \frac{\vec{f}}{m_2} = \frac{m_2 \cdot \vec{g}}{m_2} = \vec{g} \qquad (6.2.1)$$

which means that the acceleration of an object located in a gravitational field equals the magnitude of the field and does not depend upon the mass of the object. We shall return to this point when we deal with the equivalence of the inertial mass and the gravitational mass. The field $\vec{g}_1$ can vary from one field point to another. It depends upon the distance from the source (m_1), and its direction. We find that the field is a continuous entity which fills the whole space, and it varies continuously from one field point to another.

If there are several field sources, each one of them makes its own contribution according to equation 6.3, and the total field is the sum of the partial fields. In order to find what the field is at a certain point, we can use m_2 as a test particle. Let us place it in different locations in space, and, by measuring the force acting upon it, we can calculate the field at each location, and we can draw a map of the gravitational field in space. Any object introduced into this space will interact with the field at its location, and thus we have a local interaction; there is no need for an immediate interaction over distance. The interaction is between the object introduced and the field at its location.

The transference of momentum (and energy) is performed locally between the object and the field. (This description requires that the field has momentum and energy. We shall discuss this topic in detail in the next section.) The field acquires the momentum, and it transfers it with a finite speed to the other objects participating in the interaction. Hence, the objects already present in the field will feel the presence of a newly introduced object later than the moment it was introduced—by the time needed for the momentum (and the information) to propagate over the distance between them. The field acts as a mediator, which transfers interactions between masses located apart from each other.

6.3.   Momentum and Energy of a Field

We can relate momentum and energy to a field if we treat the field as a continuous medium. In order to demonstrate this, consider a continuous medium (a fluid-like medium) in which no discrete masses exist, but in which exists a continuous mass density ρ. If the fluid is moving at a velocity $\vec{u}$, we multiply ρ by the velocity to obtain a vector $\vec{h}$; $\vec{h} = \rho \cdot \vec{u}$, which is the momentum density of the fluid (in parallel to the momentum $\vec{p}$ of a particle given by: $\vec{p} = m \cdot \vec{v}$).

From the relation between mass and energy, we know that the energy density of the fluid ε, is equal to the multiplication of the fluid density by the square of light velocity: $\varepsilon = \rho \cdot c^2$. This formula is parallel to Einstein's formula $E = m \cdot c^2$. Thus we obtain the connection between the momentum density $\vec{h}$, and the propagation of energy density:

$$\vec{h} = \rho \cdot \vec{u} = \frac{\varepsilon}{c^2} \cdot \vec{u}. \tag{6.4}$$

In a static configuration $\vec{u} = 0$, we have only static energy density.

Let us use the same concept for the force field. The field is the parallel to the fluid. Its energy density is again ε. According to the STR, the presence of energy density implies the presence of mass density, $\rho = \varepsilon/c^2$. The principle difference between the fluid medium and the field is that the propagation of energy in the field goes with the speed of light, and hence the momentum density is:

$$\vec{h} = \frac{\varepsilon}{c^2} \cdot c = \frac{\varepsilon}{c} = \rho \cdot c. \tag{6.4.1}$$

Usually, when only static particles are present in space, there exists a static field emerging from these particles. In this situation there is only static energy density, and no momentum density. When an object of mass m_2 is introduced, an interaction between the field and m_2 occurs, and energy (negative or positive) is added to the field. At the beginning, this addition is local, at the location of m_2. The change in the energy of the field expands from the point at which it was formed, at the finite velocity, c. With this flow of energy a momentum density is created too. Through the interaction the particle exchanges momentum and energy with the field, and these momentum and energy flow and spread over the field, reaching other locations in the field at later times, where the time delay depends upon the distance of each location from the origin of the change. When this change reaches the location of m_1, an interaction between m_1 and the field occurs (the new field = the old field + the change in the field), and this interaction is also a local one. Through this interaction, momentum and energy are exchanged between the field and m_1. Thus, the changes in the field generated by the introduction of m_2 at its location reach m_1 at a later time. The conservation of momentum and energy is guaranteed by the local character of the interaction between the masses and the field. The field is used as a mediator which transfers the interaction, and it keeps the balance of momentum and energy during the time interval between the interactions of m_1 and m_2.

We denote the energy flux by $\vec{S}$; $\vec{S} = \varepsilon \cdot c$, and its name is "Poynting vector" (after the English scientist J. H. Poynting, who calculated this entity for electromagnetic fields in the 19th century.) The momentum density is given by: $\vec{h} = \rho \cdot c = \vec{S}/c^2$. When we locate a mass at some point, the field of the mass expands with a finite speed, so that another mass, located at that vicinity earlier, will feel the presence of the new mass after some time delay.

Thus, by relating momentum and energy to the field, we find that the interaction is not simultaneous for all the objects participating in the interaction. Instead of the law of action and reaction, we have the law of the conservation of momentum, which is a more general one. The momentum balance is kept by the field, used as a mediator which transfers momentum and energy between the interacting particles. By using the concept of field, we relate the action on the particle to the properties of space at the location of the particle. These properties are induced on space by the sources of the field. But once the field is mapped over space, we no longer need use the distribution of the sources explicitly, and we can calculate expected interactions by using the information gathered in the map.

6.4.   Potential Energy

For future use, let us define here the concept of the potential energy of the field.

The potential energy at a field point is defined as the work needed to bring a unit mass from infinity to that point. As the gravitational field is an attractive one, the work needed to bring mass from infinity is negative (the mass can perform work on other objects). The expression for the potential energy, ϕ, in a field $\vec{g} = -G(m/r^2)\hat{r}$ is:

$$\phi = -G\frac{m}{r}. \tag{6.5}$$

The potential difference between two field points, 1 and 2, is: D

$$\Delta\phi = \phi_2 - \phi_1 = -G \cdot m \left(\frac{1}{r_2} - \frac{1}{r_1}\right) = G\frac{m}{r_1 \cdot r_2}(r_2 - r_1) \tag{6.5.1}$$

When dealing with a location which is very far from the center of the source mass (r_1, r_2 large), such as the case we are in on the surface of the earth's globe, and the difference between r_1 and r_2 is relatively small, we can see expression 6.5.1 as combined of two parts: $G(m/r_1 \cdot r_2) \simeq G(m/r^2) = -\vec{g}$, and $r_2 - r_1 = \vec{l}$, which is the space difference between the two points. We find that the difference in the potential energy between the two points is:

$$\Delta\phi = -\vec{g} \cdot \vec{l}. \tag{6.5.2}$$

(Note that $\Delta\phi$ is given by a scalar product between $\vec{g}$ and $\vec{l}$, which means that only the component of $\vec{l}$ in $\vec{g}$'s direction contributes to $\Delta\phi$).

All the above considerations about the field concept can be applied to electromagnetic theory and thus define an electric field, etc. Modifications have to be made due to the fact that there exist two types of electric charges—positive and negative—while in gravitation we know only one type of charge—the mass. Another important difference between electromagnetic theory and gravitation is that the electric force depends upon the electric charge, but the acceleration caused by this force depends upon the mass of the charged object. The consequence of this is that acceleration imparted to electric charges by an electric field depends upon the ratio of charge to mass and is different from one charged particle to another.

6.5. Inertial Mass and Gravitational Mass

From equation 6.2.1 we find that the acceleration imparted to an object by a gravitational field does not depend upon the object's mass. The reason for this is that the object's mass which creates the gravitational force and the mass which creates the inertia of the object (which is the object's response to the accelerating force) are equal to each other. What are these two types of masses?

In principle we should expect the existence of two types of masses: the gravitational mass (which is the entity which creates the gravitational force,) and gravitation acts on this mass. This gravitational mass denoted by m_g fulfills the role of the electric charge in elctromagnetic theory, and it appears in equation 6.2. The other mass is the inertial mass, denoted by m_i, which is the entity responsible for the inertial properties of objects, and it appears in the equation of motion:

$$\vec{a} = \frac{\vec{f}}{m_i} = \frac{m_g \cdot \vec{g}}{m_i} \tag{6.6}$$

where $\vec{a}$, $\vec{f}$ are the acceleration and the force, and m_i is the inertial mass of the accelerated object.

We observe that the acceleration is proportional to the ratio m_g/m_i. If this ratio has the same value for all the objects in the world, we can fit the unit system in such a way that it is equal to unity. Thus the mass is factored away in equation 6.6, and the acceleration equals the field, as we obtained in equation 6.2.1. It follows that the acceleration is equal for all objects, regardless of their mass. This phenomenon always seemed questionable. Galilei performed a large number of experiments (which according to the folklore, were performed from the leaning tower of Pisa) to verify that the free-fall acceleration is indeed equal for all objects. We can formulate the problem in a trivial form: Intuitively, one assumes that as the mass of an object increases, its inertia increases too and larger force will be needed to accelerate it. Hence, it seems that heavier objects should fall slower. On the other hand, the accelerating force in a free-fall is the gravitational force which increases with the mass, and hence, for heavier objects the accelerating force is larger. If the increase in inertia and the increase in the gravitational force are exactly proportional to each other, the speed of free-fall will be equal for all objects, and this is what was found by Galilei.

In Newton's theory, the law of gravitational force has nothing to do with the equation of motion, and there is no reason to expect that a connection between the two types of masses exist. As mentioned above, we find a counter example in electromagnetic theory, where the force law is:

$$\vec{f} = q \cdot \vec{E} \tag{6.7}$$

where $\vec{E}$ is the electric field, and q is the electric charge. Substituting equation 6.7 into equation 6.6 yields the acceleration imparted to the charged particle by the electric force, and we find that it is proportional to the ratio q/m_i. It may be different for each type for charged particles. Here q plays a parallel role to that of the gravitational mass in equation 6.2, and m_i is the inertial mass appearing in equation 6.6. Comparison of equations 6.6 and 6.7 "suggests" that the gravitational mass be called "gravitational charge."

A huge amount of experimental work has been invested to verify the equality of m_i and m_g. It has been found that they are equal up to one part in 10^{11} (which is the precision limit of the experiment)[1]. As we do not believe in arbitrariness in nature, this equality tempts us to conclude that indeed there exists an intrinsic relation between the properties of matter connected with the two types of mass: inertia, which is the reaction of matter to force action, and the gravitational interaction, which appears as the attractive force between masses. This connection manifests itself through the fact that all objects have the same acceleration under the action of gravitation, and they are influenced in the same way by inertial forces.

6.6.   The Principle of Equivalence

The first move towards the creation of the GTR was the formulation of the principle of equivalence, proposed by Einstein in 1907. This principle is based on the phenomenon explained in detail in the preceding section—the equality of the inertial mass and the gravitational mass. This equality led Einstein to the conclusion that there exists an equivalence between the phenomena in which these masses are involved—the gravitational phenomena and the effects observed in accelerated systems. The principle of equivalence can be formulated loosely as: "There are no physical effects which can distinguish whether a system is accelerated with a constant acceleration or whether it is under the influence of a homogenous gravitational field."

Some examples might clarify this lengthy statement.

Let us locate our laboratory in an elevator which has an infinite track. When the elevator is at rest on the earth's surface, an observer in the elevator feels gravity by the pressure which the elevator floor exerts on his feet, keeping him from falling through. When the elevator is released, it starts moving in a free-fall motion under the influence of the gravitational field. Along a short distance this motion is a constant acceleration motion. As the observer is moving with the elevator with the same acceleration, he feels no pressure on his feet anymore; he feels as if he is in a gravity-free space. Being in a constantly accelerated system has transformed away the effects of gravitation.

Consider now the opposite case. Suppose that the elevator is indeed in a gravity-free space. When it is at rest, no force is felt by observers in the elevator. Now, let the elevator be pulled with constant acceleration upwards (with the magnitude of the acceleration equal to the free-fall one). Because of their inertia, the observers in the elevator feel an upward force exerted by the elevator floor on their feet, and the force pressure on their feet is exactly equal to the one they felt on being at rest in a gravitatonal field. If the observer releases an object in the accelerated elevator, it will continue its motion with the velocity it had at the moment it was released, while the elevator (and the observer) continue its accelerated motion upwards. The released object will seem to the observer as if it moves with a constant acceleration downwards. This is exactly the behaviour of an object in a gravitational field.

Any physical experiment performed in the accelerated elevator will show the characteristic behaviour of objects being in a gravitational field. Thus, an accelerated system in a gravitational-free space generates the same effects expected for a homogenous gravitational field.[2]

The examples given above show that when one describes the natural world, the gravitational phenomena and observations made in an accelerated system are interchangeable, without any preference for one point of view or the other. The inertial forces in a rotating system are related to gravitation in the same way. The well known example is that of a space-craft orbiting around the earth. The "weight-free" behaviour of the objects in the spacecraft is due to centrifugal force emerging from their orbital rotation, which exactly balance the earth's gravitation. In the same way, for a spacecraft far away in space, rotating the spacecraft around its axis, generates the gravity feeling for the astronauts in a gravity-free space. Any type of inertial force can be used to transform away gravitational effects (linear inertial force, centrifugal force, Coriolis force, etc.). A system whose gravitational effects were transformed away may be considered a local inertial system, and laws of nature will have the same form as they have in the STR.[3]

The principle of equivalence gives rise to interesting predictions: the bending of light trajectories in a gravitational field and the shift of wave-length of light moving in the direction of the gravitational field.

6.7.   Bending of Light Trajectories

For the illustration of the bending of light trajectories, let us return to the free-falling elevator.

Let a small hole be drilled in the side wall of the elevator, at a height h above the floor (see Figure 6.1). Let a horizontal ray enter this hole at the moment the elevator is released and reach the opposite wall. How will the trajectory of this ray seem to an observer located in the elevator and to an outside observer? The inside observer, who moves with the elevator in a free-fall, considers the elevator to be an inertial system, and he expects light to move along straight line, and hit the opposite wall also at height h above the elevator's floor. He can verify by measurements that both points, the hole through which the light entered, and the point it hits the opposite wall are at the same height. (If the measurements yield different heights for the two points, he will have a physical experiment which enables him to observe that his system is accelerated.) The outside observer will also mark the point at which the light hits the opposite wall. This event, the hit of light at a certain point on the wall, is a physical event and both observers agree about the height of the point (h) above the floor. But for the outside observer, the elevator is moving downward, and while

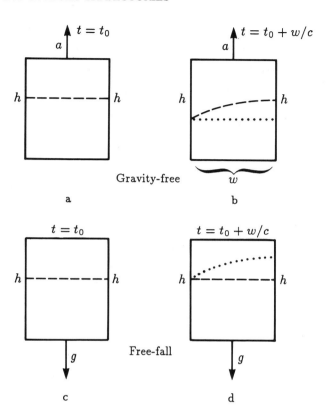

FIGURE 6.1. The path of a horizontal light ray in the elevator.

a, b.   An accelerating elevator in a gravity free space.

a.   The light enters the elevator at $t = t_o$.

b.   The dotted line is the path observed by the external observer. The dashed line is the path observed by the internal observer.

c, d.   Free falling elevator.

c.   The light enters the elevator at $t = t_o$.

d. $t = t_o + w/c$. The dashed line is the path observed by the internal observer. The dotted line is the path observed by the external observer.

the light traveled across the elevator, the opposite point h is already lower than the height it was at the moment the light first entered the elevator. The point of entrance, and the point of hitting on the opposite wall are physical events marked by the two observers. But for the inside observer the line connecting these two points is a horizontal straight line, parallel to the elevator's floor, and for the outside observer this line starts at a certain height and ends at a lower one. The outside observer finds that the light trajectory is bent downward. He must conclude that the gravitational field which caused the free-fall of the elevator also caused this bending of the trajectory of the light ray. (In the next chapter we describe the physical experiment in which this prediction was verified.)

The free-fall acceleration transformed away the gravitational effects for the inside observer, and as no gravity is felt in his system, he expects no bending of light ray in his observations. (In our considerations we neglected the STR effects assuming that the velocities are small.)

In the opposite case, the elevator accelerates in a gravity-free space. The outside observer sees that the elevator is moving upwards while light moves horizontally across the elevator. Due to this motion, he finds that the light hits the opposite wall at a point closer to the elevator's floor than the point through which it entered the elevator. The inside observer relates any deviation from the STR he observes to the existence of the gravitational field which he feels. For him, the elevator is at rest in a gravitational field, and he concludes that the light hits the opposite wall at a lower height because it was bent downward by the gravitational field.

The trajectories of light for the two cases are drawn in Figure 6.1. Denote the elevator width by w. Then the time Δt needed for the light to cross the elevator is: $\Delta t = w/c$. In Figure 6.1, a, b we draw light trajectories in the upward accelerated elevator in a gravity-free space. The dotted line marks the trajectory observed by the outside observer: The light moves in a straight line and hits the opposite wall at a lower height due to the elevator's motion. The dashed line marks the light trajectory as observed by the inside observer. For him both points marked by h are at the same height, and the light hits the opposite wall at lower point because it is bent by gravity.

The same details are drawn in Figure 6.1, c, d for the free-falling elevator in a gravitational field. For the inside observer, both h marks are at the same height, and light moves along a straight line across the elevator. The outside observer knows that at the moment the light hits the opposite wall, the h mark there is lower than it was at the entrance time, due to the downward motion of the elevator. He concludes that the light trajectory was bent due to gravity. (If we want to know exactly the form of the curvature of the light trajectory, we can repeat the same experiment many times, each time with a different width elevator, and thus obtain the height of the hitting point as function of the width.)

6.8.   Shift of Wavelength of Light

The effect of wavelength shift due to gravity may be explained in a similar way. Suppose that a light ray is emitted downward from the ceiling of an elevator, at the moment it begins its upward acceleration in a gravity-free space. A detector located on the elevator's floor detects the light a short time after it was emitted—the time needed for the light to cross the elevator's height. During this time, the elevator already gained a certain velocity, which equals the acceleration times the time interval between emission and detection. Due to the Doppler effect (see Glossary), the detector, moving towards the light, finds a blue shift in the spectrum (contraction of lightwave). This will be the explanation of the blue shift given by the outside observer (who observes the elevator's motion). But an inside observer assumes that he is at rest in a gravitational field. He concludes that the blue shift he observes is due to this gravitational field.

The same considerations take place in the opposite case—the free-falling elevator. The inside observer, who deems himself as being in an inertial system, will expect no shift in the wavelength of light. Indeed, the detector will find no such shift. The outside observer, who knows that the elevator is in a downward motion, expects a red shift to be detected, due to the motion of the elevator in the direction of the light motion. The observation that no red shift is found by the detector must be the same for both observers. Hence, the outside observer concludes that an effect due to the gravitational field compensates exactly for the red shift expected. This effect is a blue shift caused by the presence of gravity when light is moving in the direction of the field.

The magnitude of this effect can be calculated: If the light was emitted at the instant the elevator was released for the free-fall, then the elevator's velocity when the light reached the floor equal $\vec{g} \cdot \Delta t$, where $\vec{g}$ is the acceleration, and Δt is the time needed for the light to travel the elevator's

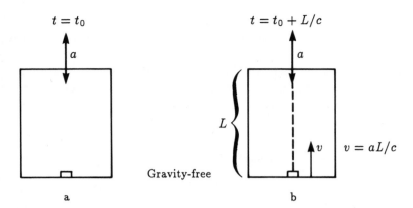

FIGURE 6.2.   The path of a vertical light ray. At $t = t_o$ the elevator starts accelerating upwards and the light is emitted downwards. The detector on the floor observes a blue shift of the light spectrum.

height.   Denote this height by $\vec{l}$: $\Delta t = \vec{l}/c$, and the elevator's velocity when the light reached the detector is: $v = \vec{g} \cdot \vec{l}/c$. Inserting this value of v into the classical expression for the Doppler effect yields:

$$\lambda_2 = \lambda_1 \left(1 + \frac{v}{c} \right) = \lambda_1 \left(1 + \frac{\vec{g} \cdot \vec{l}}{c^2} \right) \qquad (6.8)$$

where λ_1 is the wavelength of the light emitted from the ceiling, and λ_2 is the wavelength detected by the receiver.   The sign in the brackets is determined by the direction of the elevator's motion relative to the direction of the light motion.   For the shift in the wavelength, $\Delta \lambda = \lambda_2 - \lambda_1$, we obtain:

$$\frac{\Delta \lambda}{\lambda} \cong \frac{\lambda_2 - \lambda_1}{\lambda_1} = \frac{v}{c} = \frac{\vec{g} \cdot \vec{l}}{c^2}. \qquad (6.8.1)$$

The same calculations hold for the case of the accelerated elevator in gravity-free space.

In Figure 6.2 the accelerated elevator in a gravity-free space is shown. Light is emitted from the ceiling at the beginning of the acceleration and reaches the detector on the floor at a later time $\Delta t = \vec{l}/c$. The velocity of the elevator at that moment is $v = \vec{g} \cdot \vec{l}/c$. The detector on the floor discovers a blue shift of the light by the amount of $\Delta\lambda/\lambda = \vec{g} \cdot l/c^2$. The outside observer relates this shift to the elevator's motion at the moment of the arrival of the light. According to the interpretation of the inside observer, the elevator is at rest in an homogenous gravitational field, and he concludes that the blue shift is due to gravity.

To relate quantitatively the blue shift to gravity, let us use the potential energy of the gravitational field. The potential difference along the height difference $\vec{l}$ between the elevator ceiling and its floor is given in equation 6.5.2: $\Delta\phi = \vec{g} \cdot \vec{l}$. The acceleration $\vec{g}$ here is the elevator's acceleration which "generates" the gravitational effect. (An inside observer in an accelerated elevator in a gravity-free space relates to the gravitation field he "observes" the same acceleration as the one which generated the gravitational effect, in the opposite direction). Hence:

$$\frac{\Delta\lambda}{\lambda} = \frac{\vec{g} \cdot \vec{l}}{c^2} = \frac{\Delta\phi}{c^2} \tag{6.9}$$

where $\Delta\phi$ is the potential difference along the path. When light moves in the direction in which the potential decreases, it is shifted towards the blue end of the spectrum, and vice versa. Hence, light emitted by massive objects (emitted from lower potential to higher one) is shifted towards the red end of the spectrum. This may be interpreted that light loses energy (transformed into lower energy photons) when climbing up the gradient of the gravitational potential. (In this context massive objects are called 'potential wells.')

Let us calculate the red shift in light travelling from the sun to the earth. The potential difference along this path is the difference between the potential on the sun's surface and the one on the earth's surface:

$$\Delta\phi = \phi_e - \phi_s,$$

where ϕ_e is the potential on the earth's surface, and ϕ_s is the potential on the sun's surface. The potential on the earth is negligible relative to that of the sun, and hence: $\Delta\phi = -\phi_s$. Inserting numerical values we find[4]:

$$\frac{\Delta\lambda}{\lambda} = -\frac{\phi_s}{c^2} = \frac{G \cdot M}{R \cdot c^2} = 2.12 \times 10^{-6}. \tag{6.10}$$

This effect is very small. The light emitted from the sun suffers wavelength shift due to many disturbances (e.g, peculiar thermal velocities, local flow of matter due to convection), and the gravitational red shift can hardly be observed on top of all these disturbances. On the other hand, this effect can be observed clearly in light emitted by white dwarf stars. White dwarfs are stars whose mass is of the same order of magnitude as the solar mass, but their radius is smaller by a factor of hundreds. Hence the gravitational shift is larger in these stars and is observed clearly. Experiments performed on the earth's surface by Pound and Rebka (1960) with γ rays along a height of 22 meters have shown this effect clearly. Later we shall discuss this experiment in detail.

Shift in light wavelength is connected with the rate of local clocks. (Most precise measurements of time units are performed by measuring the frequency of a certain spectrum line). If we want to compare time intervals between different events, we can use the information carried by light from one event to the other. Let us discuss the following experiment: An observer located in a gravity-free space ($\phi = 0$) sends a light pulse composed of a certain number of oscillations and then measures the time length (Δt_o,) of this pulse. Another observer, located in a potential well with a potential ϕ, receives this pulse and measures its time length t, according to his clock. Because of gravity, the light moving down the potential gradient (in a potential well ϕ is negative) is blue shifted, according to the equation:

$$\lambda = \lambda_o(1 + \phi/c^2)$$

(recall that $\phi < 0$) where λ is the wavelength measured in the potential well, and λ_o is the wavelength at the zero potential.

Due to the blue shift (higher frequency), the time length of the pulse (composed of the certain number of oscillations) measured in the potential well, Δt, will be shorter than Δt_o. This means that a unit time on the clock located in the potential well includes more light oscillations than the unit time at the zero potential. Since both clocks are mechanically equal, the observer in the potential well concludes that his clock is slowed down relative to the outside clock by the factor:

$$\Delta t = \frac{\Delta t_o}{(1 + \frac{\phi}{c^2})}. \tag{6.11}$$

The topic of time dilatation has been treated here very briefly. A more rigorous derivation of time dilatation and shift of light wavelength may be found in a book by Tolman[5]. The solution to the 'Twin paradox' (which is usually mentioned in connection with the STR) lies actually in the connection between time dilatation and gravitational wavelength shift. A detailed description of this topic is given by Harpaz[6].

6.9.   Mach's Principle

An interesting explanation of the equality of gravitational and inertial masses was given by Mach. According to his explanation (known as 'Mach's principle'), the inertial properties of material objects are created by the interaction of the object mass with all the other masses in the universe (a different interaction than the regular gravitational interaction). More specifically, a material object interacts with other masses by two types of interactions. One of them is the regular gravitational interaction, which has been known for a long time. This is a static interaction formulated by Newton's law for the gravitational force, and it gives rise to the attraction between masses. The magnitude of this interaction is inversely proportional to the squared distance between the interacting objects. Hence, it falls rapidly with distance, and we feel mainly the interaction with nearby objects (e.g., the earth, the moon, the sun). In addition to this well known interaction, there exists another one between masses which is proportional to the masses and to the relative acceleration between them. This interaction is involved in the inertial phenomenon.

When accelerating a material object, all other masses remain in their original state, and the object is accelerated relative to the universe. This gives rise to a force $\vec{f}_i$, whose magnitude is the product of the object's mass and the acceleration:

$$\vec{f}_i = m_i \cdot \vec{a}. \tag{6.12}$$

This equation has the same form as Newton's equation of motion (6.6), but its meaning is different. Equation 6.6 expresses the force acting on an object whose mass is m and which acquires the acceleration $\vec{a}$. Let us call this force the 'acting force,' $\vec{f}_a$. Equation 6.12 describes the inertial force $\vec{f}_i$, by which the object opposes its acceleration (it opposes a change in its motion). In order to keep an object moving in acceleration, we have to supply an active force which will compensate for the inertial force. This active force, $\vec{f}_a$ (the one which appears in the equation of motion) must be equal to the inertial force, $\vec{f}_i$, to keep the object accelerating. We can formulate the equation of motion in the form proposed by D'Alembert:

$$\vec{f}_a - \vec{f}_i = \vec{f}_a - m_i \cdot \vec{a} = 0. \tag{6.12.1}$$

This formula may seem to be a generalization of the first law of Newton, the law of inertia. This law states: "In order to keep an object in a given state of motion, the sum of forces acting on the object must vanish." In Newton's formulation, a 'given state of motion' means inertial motions only, including rest. In the formulation of D'Alembert we also include the state of accelerated motion as a 'given state of motion.' In this case, the inertial force should be included in the balance. (The formulation of Newton did not include inertial forces—might this be the reason why some scientists prefer to call these forces 'fictitious forces'?)

Equation 6.12 describes the situation as observed by an observer in the accelerated system. For him, the accelerated object is at rest, and in order to keep it in this state of motion (at rest in an accelerated system), the sum of all the forces it feels must vanish. This sum includes both the accelerating force and the inertial force. The formulation of D'Alembert of the second law of Newton transforms this law into the law of inertia in accelerated systems.

We may find more meaning in the equality of the inertial and gravitational masses. This equality can be written as:

$$m_i = I \cdot m_g,$$

where I is some constant to be determined later. The inertial force exerted by an accelerated object is $m_i \cdot \vec{a}$. Thus, m_i should include all the factors which participate in inertial interaction, except for the acceleration. According to Mach's approach, all the other masses in the universe participate in this interaction and must appear in m_i. But the influence of each mass should depend somehow on its distance from the accelerated object, in a way we do not know yet. Hence, m_i should include a sum of products of the masses times some function of the distance, $m_j \cdot f(r_j)$, where j is a dummy index of the other masses which participate in the interaction, r_j is the distance of the jth mass, and $f(r)$ is an unknown function of r. The sum of the products should be summed for the total number of masses in the universe, N (except for the accelerated one). Hence m_i includes: the mass of the accelerated object m_g, the sum of the products mentioned above, and some constant of interaction K. Let us write:

$$m_i = K \cdot m_g \cdot \sum_{j=1}^{N} (m_j \cdot f(r_j)) = I \cdot m_g. \qquad (6.13)$$

The equality of the inertial and gravitational masses demands that we choose a unit system in which:

$$I = K \cdot \sum_{j=1}^{N} (m_j \cdot f(r_j)) = 1. \qquad (6.14)$$

We obtain a condition for the constant K:

$$K = \frac{1}{\sum_{j=1}^{N} (m_j \cdot f(r_j))}$$

There are two factors in K which can be time dependent: r_j and N. We shall discuss this point later when we deal with the evolution and expansion of the universe.

The consequences of Mach's principle are far-reaching, much more than can be understood at first glance. According to Newton, we can find directly which system is accelerated (thus giving acceleration an absolute meaning). This can be done by a simple experiment, called by Newton "the water bucket experiment." The water surface in a bucket is flat while the bucket is at rest or in an inertial motion—(with no acceleration). When the bucket is accelerated with a linear acceleration, however, the surface of the water will be inclined at an angle which can be calculated by a vector addition of the linear acceleration and the gravitational acceleration (see Figure 6.3).

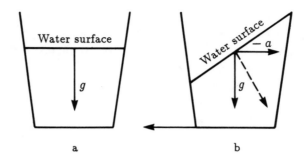

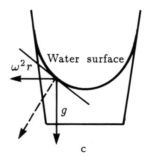

FIGURE 6.3.   The water bucket experiment.
  a.   The water bucket is at rest.
  b.   The water bucket moves with a constant acceleration a.
  c.   The water bucket rotates around a vertical axis. The water surface takes parabolic shape.

Hence, if the water surface is inclined relative to the horizon, it shows that the bucket is accelerated. If we rotate the bucket around a vertical axis passing through its center, the water will take a parabolic shape due to the centrifugal force.

Newton assumed that this is a crucial test which determines uniquely whether a system is accelerated or not. On the contrary, Mach's principle states that if the bucket is at rest, and the whole universe rotates around it, the water will take the same parabolic shape (due to the interaction with the rotating masses of the universe) as if the bucket rotates and the universe is at rest. According to Mach, the water bucket experiment will show up in any case in which relative acceleration between systems exists, and it does not distinguish which one of the systems is absolutely accelerated.

Another consequence of Mach's principle is that the inertial properties of objects exist only in the presence of other masses in the universe. A single particle existing alone in the universe has no inertial property because there is no frame of reference to which the acceleration can be related. The addition of another particle gives meaning to the inertial properties of the first particle because it serves as a reference frame for measuring acceleration. Further, the additional particle(s) is (are) the source for the interaction which appear as the inertial properties of the particle. But the amount of the other masses determines also the magnitude of the inertial interaction. The addition of more masses will add more sources for the interaction, so that the inertial properties of a material object should be proportional to the amount of mass in the universe.

How does this interaction depend on distance?

The matter distribution in our close vicinity is very anisotropic (the solar system, the Galaxy, etc.). This anisotropy of the nearby matter causes anisotropy in the gravitational attraction we feel (the attraction towards the earth's center, the attraction of the earth towards the sun, etc.). Yet we see no preferred direction for the inertial properties, and this interaction seems to be isotropic. We have to conclude that the inertial interaction depends mostly on the remote masses in the universe, whose distribution is assumed to be isotropic. The gravitational attraction falls as $1/r^2$. Hence the gravitational force exerted by remote masses is very weak although they sum up to a huge amount of mass. If the inertial interaction depends mainly on remote masses, its dependence on distance, $f(r)$ should fall less steep than $1/r^2$. The further we move from the accelerated object, the more masses are included in the sum we calculate for the interaction, but their distance is bigger. The fall in the magnitude of the interaction with distance should be such that the increase with distance of the amount of mass included in the interaction will overcome the fall of the magnitude of the interaction due to the distance. Sciama has shown in his book[7] that a dependence of the interaction on distance of the type of $1/r$, is suitable to create such a situation in which the remote masses will dominate the interaction.

It should be emphasized that this interaction is proportional to the magnitude of the acceleration: it depends on the relative motion of the object. Curious as it appears, this interaction is not the only example of an interaction depending on the relative motion of the interacting objects. In electrodynamics, we have a static interaction between charges expressed by Coulomb's law, which falls with the square of the distance. In addition, there is the magnetic interaction between charges which depends both on the distance between the charges and on their relative motion. The same entity, the electric charge, serves as the object of both interactions. In parallel, we know the static gravitational interaction, which falls with the squared distance. With Mach's principle another interaction is added which depends on a relative motion, namely the relative acceleration between the interacting objects. The same entity, the mass ($m_g = m_i$), serves as the object for both interactions and is actually the 'charge' for these interactions. The idea that both types of phenomena (gravity and inertia) are interactions between masses explains quite reasonably why the two types of masses should be equal.

There are some experimental predictions as a consequence of Mach's principle which might serve as a test for the theory. One is the prediction that a rotating hollow spherical mass should give rise to a centrifugal force near its center (which is at rest). The effects expected are very minute and no accurate equipment exists to date which can establish this effect. From experimental measurements we know that the equality of the two types of masses is accurate up to a very high degree.

6.10.   Constraints on the Trajectories of Motion

By now we have realized part of our scheme. We have found that gravitational effects and observations carried in accelerated systems are interchangeable and can be treated on an equal footing. Now we have to find

how these phenomena can be interconnected with the geometrical properties of space. We have seen in the example of the rotating disc (Chapter 5)
that an accelerated system gives rise to curved space characteristics. The
acceleration in that system was a centripetal one (center-directed acceleration). Recall that the Lorentz transformations of the STR can be described
as pseudo-rotation which is expressed in the change of the angle between
the new system axes, relative to the angle between the axes of the original
system. Still the new system is also flat, and the axes in both systems are
straight lines. Later we shall learn that the transformations of the GTR can
cause curving of the axes of the system, in addition to the pseudo-rotation.

Let us study this through a simple example:

An observer on a moving train can observe events occurring outside the
train while assuming that his system (the train) is the rest system. While
the train moves with a constant velocity $\vec{u}$ (along a straight line), all the
objects at rest in the external system will seem to him as moving with
constant velocity $-\vec{u}$ (in the opposite direction to the train motion). An
object moving with constant velocity $\vec{v}$ in a direction perpendicular to the
train motion will seem to be moving along an inclined straight line, whose
angle α relative to the train is:

$$\tan \alpha = \frac{\vec{v}}{\vec{u}}. \tag{6.15}$$

This equation is the nonrelativistic law for addition of velocities.

Suppose now that the train moves with a constant acceleration. The
velocity $-\vec{u}$, the velocity of the external objects relative to the train, will
increase with time. The composite motion of an object moving with
constant velocity perpendicular to the train will seem to the observer in
the train as a motion in a parabolic trajectory. The parabolic motion
is created by the composition (vector addition) of the constant velocity
perpendicular to the train, and the accelerated motion in the anti-parallel
direction to the train (as with the trajectory of an object in a horizontal
throw, equation 5.23). Most objects in the external system (which move
with different velocities and directions) will seem to the observer in the
train as performing motions composed of their original motions in the
external system plus the parabolic component created by the train's acceleration (the rest system of the observer). The train observer may try
to explain the trajectories he observes. If he knows all the factors which

caused the motions of the objects relative to the external system, he will have to invent a special additional force which forces on all objects the parabolic constituent of their motion. Alternatively, instead of dealing with forces he might suggest that there is some constraint which forces all objects to follow the parabolic path. (The fact that while going from Boston to London we have to follow the curvature of the earth's surface instead of moving along a straight line drilled through the earth is also such a constraint). In the same way as in the case of the inertial forces on the rotating disc, the parabolic nature of the trajectories does not depend on the particular mass of the objects. This means that the fictitious force which created the parabolic component of the motion should be proportional to the mass (hence it is factored out from the equation of motion). Inertial forces are always proportional to the mass of the object they are acting on.

The train observer may find a more "clever" explanation. Instead of inventing fictitious forces or thinking about constraints, he might "invent" a system of reference in which the parabolic component of motion will be transformed away. Such a system is one which is in a constant acceleration relative to the train, and which is directed towards the rear end of the train. (We know in advance that such a system exists, namely the system external to the train.)

According to the principle of equivalence, we know that if the train is at rest and a homogenous gravitational field acts in the direction of its rear end, all the phenomena in the external system will seem to the observer in the train exactly the same as in the case of the accelerated train. The train observer may choose what interpretation he wants to give to his observations: either by attributing them to the existence of an external gravitational field or by attributing them to the acceleration of his own system. In both interpretations, the parabolic component of the motion of objects is fundamental.

Another interpretation is possible, too, an interpretation which will deal with the phenomena only without going into the factors which caused the phenomena. (Often we are obliged to do so when we know very little about the causes of phenomena in complicated situations in nature, but still we want to analyze the phenomena and develop predictions about their evolution in time). According to this interpretation, the parabolic behaviour of the objects is due to the space they are located in. The fact that all the objects, regardless of their mass, perform the same type of trajectories encourages us to choose this interpretation.

To better understand the meaning of constraints, let us look at some simple examples in which constraints determine the characteristic motions of objects.

Consider a bead on a closed loop of a wire. The only trajectories allowed for the bead are limited to motions along the loop, irrespective of the forces that might act on it. Let us follow the bead's motion and try to conclude from our observations what forces acted upon it. This can be done, provided that the constraint on the bead's motion is always remembered. This constraint is the same for any bead on the loop, regardless of its mass. If a force acts on the bead, its response will be according to the constraint. The component of the force directed along the loop will influence the bead's motion and impart to it an acceleration according to the equation of motion since this is the direction the bead is allowed to move. The components of the force perpendicular to the direction of the loop will contribute nothing to the bead's motion. The reaction force of the loop will compensate for their action.

Calculating the motion of the bead as a function of the forces acting on it can be done in two different ways. First by taking account of the constraint, we include in our calculations only those components of the forces which contribute to the motion along the allowed trajectory and ignore any interaction between the external forces and the loop. Second, include in the calculations the full action of the external forces, together with the reaction forces of the loop to these forces. The net result will be the same. But we shall find that the reaction forces exerted by the loop do not depend on the specific properties of the loop (subject to the condition that the loop does not break). The force exerted by the loop is a reaction force, hence it will always be equal to the force needed to keep the bead in its allowed trajectories.

Thus, the additional information concerning the external forces in full add nothing, either to our ability to predict the bead's behaviour or to our insight into the phenomena. Both interpretations given above for the behaviour of the bead are legitimate—either a constraint which allows certain trajectories only or a balance of forces which determines the bead's motion at each moment. Since the facts observed according to the two interpretations are the same, there is no observation (of the bead's motion) which can lead us to prefer one interpretation over the other. Usually, the constraints' interpretation is more convenient to handle.

As a second example, let us consider the trajectories of human beings on the earth's surface (prior to the age of planes and spacecrafts). Any human motion (including ships routes, communication lines, war movements or recreational travels) was confined to the surface of the earth. An external observer, watching and studying our motions, would conclude that the space available for human motions was limited to the surface of the globe. He might conclude that there exists a constraint which forces human being to move on the globe surface only. To explain this phenomenon by consideration of the actions of forces, he might invent a system of forces, a system in which one of them might be an attractive force towards the center of the globe and opposed by other forces which prevent humans from being immersed into the globe. Such a force system might explain the limitation of the motion space to the surface of the sphere only. But, for an external observer, the interpretation that a certain constraint limits the motions of human beings to the surface of the sphere might be more compact and convenient for use.

In both examples given above, the constrains may be described as a curved subspace embedded in a more general space. The example of the bead deals with a one-dimensional space in which the bead "lives," and this space (the wire loop) includes all the locations available for the bead. This space is a curved, closed and infinite space, as the bead can repeat its trajectories along the loop an infinite number of times without reaching an endpoint of the space. The meaning of closed space is that it includes all the points allowed for the bead. By reducing the subspace of the bead to one dimension only, we did not lose generality nor any physical information concerning the bead.

The same considerations hold for the human beings on the surface of the earth. This surface is a two-dimensional space (closed, infinite and curved) which is a subspace of the three-dimensional space. It is closed because it includes all the motions allowed for the human beings. Through their motions they will never reach an endpoint of their space. The curvature of this space can be observed directly by observing it from a higher dimensional space (like a picture taken from a spacecraft) or by measuring geometrical objects in the subspace (the surface), as was shown in Chapter 4.

For the sake of convenience we limited our description to the spatial coordinates only. Since both cases described above are time dependent, the inclusion of a time coordinate will give a more complete picture of the evolving events: a two-dimensional space of length and time for the bead and a three-dimensional space of surface and time for the human beings.

Determination of the nature of the curved subspace which suits the constraints can be performed either by calculating the forces which generate the constraints or by observing the motions of objects subject to the constraints. In both ways—once the nature of the subspace is determined—further calculations and predictions become simpler by using the properties of the space as the basis for the calculations. The description of the properties of the general space, or its geometry, is equivalent to the description of the subspaces in which objects' motions can take place. Knowing the geometry of the space enables us to know what trajectories are allowed for objects in this space.

The method of the GTR is quite similar to the method described above. All forces acting on objects whose magnitude is proportional to the objects' masses can fit into this method because the acceleration they impart to the objects does not depend on the object's mass. This class of forces includes the gravitational and intertial forces. According to Mach, they are merely two types of interactions between masses. Let us call this class of forces general gravitational forces.

From these forces, a general gravitational field can be determined. This field serves as a source for the constraints, determining which is the allowed space for the motions of objects. This space can be described by its geometry. Hence, the description of the general gravitational effects (Chapter 5, section 8) and their influence on objects' motions are given by the geometrical description of space, where in the geometrical description we include the constraints acting in the space. The calculation of the gravitational and intertial forces is replaced by this geometrical description. Since the sources for the the general gravitational forces are the masses which exist in space, the geometry of space is determined by the distribution of the masses in the universe. This topic—determining the geometry of space by the distribution of masses in the universe—is the core of the GTR, and we devote the rest of this book to studying this topic.

Some limitations must be imposed on the analogy between the GTR method and the method connected with the simpler cases of the bead and the spherical surface. In those simple examples, the constraints determined a lower dimensional subspace than the general space in which the forces (the sources of the constraints) acted. In the GTR, the four-dimensional spacetime of nature is both the arena of the forces generating the constraints and the constrained space determined by them. In the real three-dimensional space of the GTR, each object can reach any location in space; the constraints determine the types of trajectory allowed. But in the four-dimensional space (in which each point represents an event), each point includes the information about the location of the event in the three-dimensional space and the information about the time it occurred. The points in the four-dimensional space which a certain object can reach depends on its three-dimensional location and on the velocity it is moving in. It should be clear that the number of events in which a material object can participate is very small in respect to the total number of events occurring in the universe. The limitation emerges from the fact that there are some world trajectories allowed for an object but many are not allowed. The allowed trajectories are determined by the system of constraints of the space.

We conclude that in the GTR we work with a four-dimensional spacetime in which the general gravitational fields were transformed away. Instead of these fields, we have a curved space, whose curvature reflects the constraints imposed on material objects by these fields. The geometry of space, which includes the description of its curvature, includes all the information about the general gravitational fields which is relevant for the description of objects' motions in this space. The goal of the GTR—to include the description of accelerated systems in the general theory—is fulfilled through the description of a curved space. This description also suits the description of the gravitational phenomena. What is left now is to see how the geometry of space and its curvature are determined by the general gravitational fields. The geometry generated by the distribution of the matter in the universe creates motion in the matter, and thus creates changes in the same distribution. This topic is dealt with in the next chapter where we deal with Einstein's equations.

Comments and References

1.   Roll, P. G., Krotkov, R. "The Equivalence of Intertial and Passive Gravitational Mass," *Ann. Phys.,* (1964) **26,** 442.

2.   The equivalence is between systems accelerated with a constant acceleration and homogeneous gravitational fields. Effects which are dif-

ferent from those which appear in accelerated systems will appear in non-homogeneous gravitational field. For example, two objects falling in "parallel" in a gravitational fields generated by a point source are moving towards the center of gravity. They will gradually come nearer to each other, while in an accelerated system they will move in really parallel trajectories.

3. Moller, C. *Theory of Relativity.* Oxford University Press, London, 1966.

4. Adler, R., Bazin, M., Schiffer, M. *Introduction to General Relativity.* McGraw-Hill Book Co., New York, 1965.

5. Tolman, R. C. *Relativity, Thermodynamics and Cosmology.* Oxford University Press, London, 1962.

6. Harpaz, A. "The Twin Paradox and the Principle of Equivalence," *Europ. J. Phys.,* 1990, **11,** 82.

7. Sciama, D. W. *The Unity of the Universe.* Anchor Books, Doubleday Corp., New York, 1961.

Einstein's Equations

Einstein's equations describe the connection between the distribution of matter and energy in space and the curvature of that space, i.e., how matter distribution generates space properties.

7.1.   A Geodesic Line

Before going into the details of the equations, let us introduce an important concept—the geodesic line (or the geodesic). The simple definition of a geodesic is: "a shortest line connecting two points." More generally, the geodesic is the extremal line connecting two given points (where an extremal is the shortest or the longest line). The meaning of the longest line connecting two points is not always clear, so we shall treat the shortest line only. The physical meaning of the geodesic line is: "The world line of free particle" which means the path in a four-dimensional space taken by a particle on which no forces are acting except for the geometrical constraints created by the the general relativistic effects. A free fall is an example of such a path.

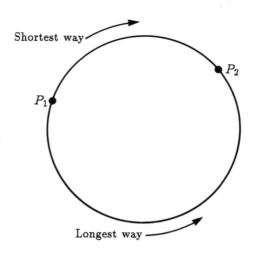

FIGURE 7.1.   A bead on a closed loop. Two extermal trajectories between P_1 and P_2. One of them is the shortest, and the other is the longest.

Let us discuss first the geometrical meaning of a geodesic in a three-dimensional space. It is clear that in a flat space, the shortest line connecting two points is a straight line. However, in a curved space, a geometrical straight line is not always possible and the geodesic could be a curved line. This definition is demonstrated clearly in the two examples of constrained spaces given in the preceding chapter, namely: the bead on the closed loop and human beings on the surface of the earth. The curved space of the bead includes one allowed path only—a motion along the loop. Hence, in order to move from P_1 to P_2 (see Figure 7.1), the bead must follow this path. Note that there are two ways to acomplish this: the shortest line and the longest one. These two lines are extremal. Thus, defining a geodesic as the extremal line includes both possibilities.

In the example of human beings on the earth's surface, there are infinite possible paths to move from point P_1 to P_2, where the shortest line connecting the two points is along a great circle of the sphere passing through these points. This line is an extremal line. Moving along the great circle in the opposite direction will bring us also from P_1 to P_2 along a longer line (which is also an extremal line), but not necessarily the longest line connecting the two points. When an object is constrained to move on the surface of a sphere and it moves without friction, it will move along great circles on the sphere. If it starts with a given initial velocity, its position on the great circle can be calculated for each moment. Its equation of the trajectory does not involve time, and the additional information about the time dependence is given only if the trajectory is described in a four-dimensional system.

The four-dimensional equation of the trajectory includes also the time points at which the object passes through each spatial point. Two objects, moving along the same three-dimensional trajectory with different velocities, will have different four-dimensional trajectories.

A geodesic on a surface can be determined as follows: Let us connect the points P_1 and P_2 by a flexible string and stretch it. If there is no friction and the string is free to find its location "at will", it will stabilize along the shortest line connecting the two points. If the surface is a part of a sphere's surface we can calculate the forces acting on it: These forces acting on a short segment of the string ΔS are a pair of stretching forces acting on the two endpoints of the segment in opposite directions, and the reaction force of the sphere surface N, which acts perpendicular to the surface and prevents the string from being immersed into the sphere. In a stable equilibrium, these three forces should balance each other (their vectorial addition should vanish). Hence, N must be in the plane which includes also the pair of stretching forces. The components of the stretching forces along the string direction balance each other, and their vertical components must be balanced by N. The plane which includes all the forces is the "osculating plane" (see Chapter 4), and each segment of the string is included in the osculating plane. The osculating plane passes through the center of the sphere, and its intersection with the sphere surface forms the great circle. Hence, the string which is included in the osculating plane is on the great circle. By such a method a geodesic is determined by considering static configuration. When the curved surface is not a surface of a sphere, its curvature may not be constant over the whole surface. The local osculating planes at each point will differ in direction from each other, and there is no meaning to a great circle on such a surface. However, at each small segment of the line, the geodesic is along a great circle of a spherical surface which coincides with the real surface at that point.[1]

The geometry of a geodesic in a four-dimensional space includes physics. As mentioned in Chapter 5 (section 1), the trajectory of an object in a four-dimensional space is determined by the velocity too. Hence, a free-fall is described by a curved line. This line is a geodesic, and its curvature reflects the space curvature. At this point we see how the geometrical picture of the gravitational effects leads to the motions of objects determined by these effects—a general gravitational effect generates the curvature of space. This curvature creates the constraints on motions of free objects and causes the fact that the geodesics, which are the trajectories of free objects, are curved lines in the four-dimensional space. We shall see later that we can calculate geodesics from the information we have about the curvature of space. Geodesics may also be found empirically, by following the trajectories of free objects. In a flat space, these objects will move along straight lines, and in curved space they will follow the curvature of space.

The method of calculating the extremal line is known as the "calculus of variations." We will now describe this method briefly, and start our description with a three-dimensional coordinate system.

Suppose that L is some function of the coordinates and their derivatives. We calculate the value of the integral of this function along a line connecting the points P_1, P_2. The value of the integral will depend on the specific route chosen as the integration line connecting the points. Choosing different routes for the integration yields different values for the integral. Among all the values we can calculate by integration, we seek for the route which yields the extremal (minimal) value for the integral. In order to verify that the line chosen will indeed yield the extremal value, we may try to vary the integration route by small variations, keeping fixed the values of L at the endpoints P_1, P_2. When we find an integration route for which small variations of the route do not change the value of the integral, then that line is the extremal one. Mathematical formulation of this statement is:

$$\delta \int_{P_1}^{P_2} L \cdot d\lambda = 0 \qquad (7.1)$$

where δ denotes variation, and λ is some parameter which parametrizes the line uniformly. The concept of 'parametrized line' requires some clarification. When we move along a certain line, we can record all the points through which the line passes. A list of these points, ordered in consecutive order, characterizes the line (the ensemble of the points generates the line). But this characterization is not unique. To obtain a unique characterization of the line, we have to measure the distance between consecutive points and add the distances to the list. A consecutive list of the points along the line, plus the distance of each point from the preceding one (or its distance from the origin of the measurements) characterizes the line uniquely. This line is parametrized uniformly. In this example, the distances of the points from the origin are the parameter of the line. They give a measure to the relative location of each point along the line. In a line characterized uniquely by a parameter, we have the information about the order of the points along the line, and we also have a quantitative measure for their relative location. Other functions may serve as parameters of the line, provided that they are continuous functions and they are connected to the distance parameter by a linear transformation.

The method of 'calculus of the variations' is similar to finding the minimum or maximum (extremum) of a function by finding the point at which its derivative vanishes. (The method of variations is more complicated since the function L depends upon several coordinates and their derivatives. We shall return to this question later.)

7.2.   Fermat Principle

An example of the use of this method is the 'Fermat principle' for calculation of the trajectory of a light ray. According to this principle, the trajectory of a light ray passing between two points P_1, P_2 is such that the time needed for the light motion between the two points is minimal. In the present case, the entity which we vary is the integral of the time intervals $dt = (n/c)dl$, where n is the refraction index of the medium, c/n is the velocity of light in that medium, and dl is the line interval covered by light during the time interval dt. In Figure 7.2 we describe a case in which the refraction index changes abruptly on the boundary plane between the two mediae. In this figure, l_1 and l_2 are the lengths of the light paths in medium 1 and medium 2, and n_1, n_2 are the refraction indexes of these mediae, respectively. In the case that the media over the whole space are homogenous ($n_1 = n_2$), then the velocity of light will be the same along the entire path, and the minimal time needed for light to travel from P_1 to P_2 is obtained by a motion along a straight line. If $n_1 \neq n_2$, the path chosen should be such that a longer way is travelled in the higher velocity medium and a shorter way is travelled in the lower velocity medium. The

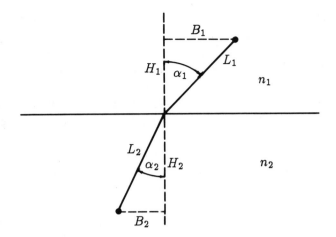

FIGURE 7.2.   Light trajectory between P_1 and P_2. The refraction index $n_2 > n_1$. Hence the light will move along a longer path through media 1.

condition is not minimalization of $l = l_1 + l_2$, but of the time, t:

$$t = t_1 + t_2 = \frac{n_1 \cdot l_1}{c} + \frac{n_2 \cdot l_2}{c}. \tag{7.2}$$

From this condition the equation of the trajectory can be calculated. Using Figure 7.2, let us decompose the distance between the two points to a parallel component, b, to the boundary plane between the two mediae, and to an orthogonal component, h, to this boundary. For the given two points, h and b are given, and b_1, h_1 are the parts of these components b and h in medium 1, etc. and b_2 and h_2 are given too. The ratio of b_1 to b_2 may change provided that their sum remains constant: $b_1 + b_2 = b$. The angles α_1, α_2 are the angles between the light trajectories in each medium and the normal to the boundary plane. Let us substitute in equation 7.2:

$$l_1 = \sqrt{h_1^2 + b_1^2};$$
$$l_2 = \sqrt{h_2^2 + b_2^2} = \sqrt{h_2^2 + (b - b_1)^2}$$

and we find:

$$ct = n_1 \cdot \sqrt{h_1^2 + b_1^2} + n_2 \cdot \sqrt{h_2^2 + (b - b_1)^2}.$$

In this expression, most of the terms are constants, except for b_1. Let us differentiate t with respect to b_1 and equate the derivative to zero. This yields an equation for b_1 which yields a minimal t:

$$c\frac{dt}{db_1} = \frac{n_1 \cdot b_1}{\sqrt{h_1^2 + b_1^2}} - \frac{n_2 \cdot (b - b_1)}{\sqrt{h_2^2 + (b - b_1)^2}} = 0$$

on substituting:

$$\sin \alpha_1 = \frac{b_1}{\sqrt{h_1^2 + b_1^2}}; \quad \sin \alpha_2 = \frac{b - b_1}{\sqrt{h_2^2 + (b - b_1)^2}}$$

we find:

$$n_1 \sin \alpha_1 = n_2 \sin \alpha_2$$

which is Snell's law for refraction. In the case where the refraction index varies continuousely (e.g., changing temperature at different heights in the atmosphere), the integration should be performed on the quantity $dt = n \cdot dl/c$:

$$t = \int_{P_1}^{P_2} \frac{n(l)dl}{c}$$

and this integration yields the path of light ray which is curved to conform to the minimal time requirement.

7.3.   The Method of Variations

An important example of the use of this method is found in mechanics, where the integral of the function L is called "the action," and L in this case equals to the kinetic energy minus the potential energy. The parameter characterizing the motion is the time. It is found that the path "chosen" by a particle for its motion from the point P_1 to P_2 is the path obtained by the requirement that the "action" along the path is minimal, or that the variation of the integral of L will vanish:

$$\delta \int_{P_1}^{P_2} L \cdot d\lambda = 0.$$

A detailed explanation of this calculation is given in the book "Lectures on Physics" by Feynman[2]. In order to calculate the variation of L due to small variations of the path, we calculate the variations of L due to small variations of the coordinate x and also the variations of L due to small variations of the derivatives of x with respect to λ, x. (Here x stands for any coordinate, and the variations should actually be performed for each one of the relevant coordinates.) If we choose the time as the parameter, then x is the velocity. Performing all the mathematics, the variation of L is equal to the sum of the derivatives of L with respect to x times δx, plus the derivatives of L with respect to x times δx:

$$\delta L = \frac{\partial L}{\partial x} \cdot \delta x + \frac{\partial L}{\partial x} \cdot \delta x.$$

Performing these calculations, subject to the requirement that the variation of the integral vanishes, we obtain the equations of motion for the particle. The advantage of this method for the calculation of the equations of motion instead of directly using Newton's equations of motion (which include the forces) is that the information about the energies in the system and their dependence upon the coordinates and upon the velocities is enough for the calculation of the equations of motion. For cases in which the description of the forces in the system is complicated (or sometimes impossible), the use of energies is generally more convenient, and the calculations through this method become simpler.

Using this method to calculate geodesics, we choose for L the interval itself, or the squared interval. (In some books on this topic the authors use the integration on the quantity $dS = \sqrt{dS^2}$. It is clear that the extremum of dS^2 is obtained at the same point where dS has an extremum. Hence, the choice between the two possibilities is a matter of convenience.)

We require that the variations of the integral on L along the path of a free particle in a four-dimensional space vanish. What quantity shall we choose for λ, the line parameter? We want this choice to be independent of the choice of the coordinate system. Hence, we should choose for the parameter a quantity which is invariant under the transformations of the coordinate system. The proper time τ (mentioned in Chapter 5, section 2) is such an invariant, hence choosing τ for the line parameter might be convenient, and the path of integration is the world line of an object whose proper time is τ. The points recorded along the path are the events through which this world line passes. Dividing the squared interval by $d\tau^2$, we obtain for the action $(dS/d\tau)^2$. Insert for dS^2 the explicit expression:

$$dS^2 = A_{ij} \cdot dx_i \cdot dx_j$$

where i, j take the values 1 to 4, and we obtain:

$$\delta \int_{\tau_1}^{\tau_2} \left(\frac{dS}{d\tau}\right)^2 d\tau = \delta \int_{\tau_1}^{\tau_2} \left(A_{ij}\frac{dx_i}{d\tau}\frac{dx_j}{d\tau}\right) d\tau = \delta \int_{\tau_1}^{\tau_2} (A_{ij}\dot{x}_i\dot{x}_j) d\tau$$

$$(7.3)$$

where A_{ij} are the components of the metric tensor, and $\dot{x}_i = dx_i/d\tau$. As mentioned above, we have to take derivatives of the integrand with respect to x_i and with respect to $\dot{x}_i$. Without going into the details of the calculations, we note that the $\dot{x}_i$ terms do not depend explicitly upon the coordinates. Hence the derivative of the integrand 7.3 with respect to the spatial coordinates will not vanish only if the components of the metric tensor depend explicitly upon the spatial coordinates (and this is what happens in curved space).

The end results of the differentiation of 7.3 are the equations of the geodesics. The number of equations obtained is equal to the number of the coordinates, where the form of each equation is:

$$\frac{d^2 x_i}{d\tau^2} + \sum_j \sum_k \Gamma^i_{jk}\dot{x}_j\dot{x}_k = 0$$

$$(7.4)$$

Here i is the index of the coordinate for which this equation holds. The first term in this equation is the second derivative of the coordinate x_i with respect to τ. The second term includes a product of the first derivatives of the coordinates with respect to τ (as they already appeared in L) times a quantity denoted by Γ.[3] This quantity is called connections or Christoffel symbols (after E. B. Christoffel who introduced them in 1869), and they include first derivatives of the components of the metric tensor with respect to the coordinates. These derivatives appear as a result of differentiating L with respect to the coordinates. The explicit form of Christoffel symbols is:

$$\Gamma^i_{jk} = \frac{1}{2} A^{im} \left(\frac{\partial A_{jm}}{\partial x_k} + \frac{\partial A_{mk}}{\partial x_j} - \frac{\partial A_{jk}}{\partial x_m} \right) \qquad (7.4.1)$$

where $\partial A_{ij} / \partial x_k$ is the partial derivative of the component A_{ij} with respect to the coordinate x_k. A^{im} is a component of the matrix $\overline{\mathbf{A}}$, the inverse matrix of $\mathbf{A}$ ($\mathbf{A} \cdot \overline{\mathbf{A}} = \mathbf{I}$). In equation 7.4.1 we observe that Γ includes products of the components of the tensor A^{im} by the derivatives of the components of the metric tensor. Hence, an equation which includes Γ will be a non-linear equation in the metric tensor components. We shall discuss later this point. What does equation 7.4 means? In the case of a flat space, the components of the metric tensor are constants, and their derivatives vanish. In such a case, the Γ terms are zero and from equation 7.4 we are left with:

$$\frac{d^2 x_i}{d\tau^2} = 0$$

which is equivalent to the Newtonian equation of motion for a free particle. Actually this is the law of inertia. The second derivative of a coordinate with respect to time is the component of the acceleration in the direction of that coordinate. When the acceleration vanishes, the object moves with a constant velocity along a straight line. But in a curved space the components of the metric tensor are functions of the coordinates, and their derivatives with respect to the coordinates do not vanish. Hence the Γ terms in equation 7.4 are not zero, and they cause deviations of the trajectory from a straight line. Actually, these terms represent the constraints imposed on the object's motions by the curvature. These constraints cause the object to follow curved trajectories through its motion.

The Γ terms we just dealt with are used also for taking derivatives in a curved space. When we differentiate vectors (and tensors) in a curved space, the components of the metric tensor should be differentiated too. Such a differentiation is called covariant differentiation. In general, a derivative of a vector will be written:

$$D \cdot B = B' + \Gamma \cdot B \qquad (7.5)$$

In this equation D represents the covariant differentiation, B is the vector to be differentiated, B' is the regular derivative of B (with the respect to all the coordinates), and Γ includes the derivatives of the components of the metric tensor. The reason for the form of equation 7.5 is that, on differentiating a vector, we move it by a small (infinitesimal) displacement from its original position and calculate by how much the vector changed through this displacement. When doing it in a curved space, we must remember that in the new position of the vector (where it is located after the infinitesimal displacement), the metric tensor is different from what it was in the original position. The change in the metric tensor due to the displacement appears in the Γ terms, which include the first derivatives of the metric tensor components with respect to the coordinates. Evidently, in a flat space $\Gamma = 0$ and $D \cdot B = B'$. Differentiating a vector twice (to obtain the second derivatives) means acting with D on the expression which appear on the right wing of equation 7.5. we find:

$$\begin{aligned}
D^2 B = D \cdot (B' + \Gamma B) &= B'' + \Gamma B' + D \cdot (\Gamma B) \\
&= B'' + \Gamma' B + 2 \cdot \Gamma B' + \Gamma \cdot \Gamma B
\end{aligned} \qquad (7.6)$$

Recall that Γ includes first derivatives of the components of the metric tensor. Hence, Γ' includes second derivatives of the components of the metric tensor. Further, the term which includes $\Gamma \cdot \Gamma$ includes products of the first derivatives of the components of the metric tensor. In curved space, whenever second derivatives appears, we also find products of the first derivatives of the metric tensor components.

7.4.   Gravitational Potentials

The role of the components of the metric tensor in equation 7.4 gives us
a clue to their role in the GTR. We require that the equations of the GTR
will correspond to the classical Newtonian equations when we deal with
phenomena connected with low velocities and weak gravitational fields.
This requirement constitutes a correspondence principle (between the GTR
and Newtonian mechanics) similar to the correspondence principle of Bohr
in quantum mechanics. In Newton's equation of motion we usually have
an acceleration term (the second derivative of the coordinate with respect
to time) and the other terms of the equation represent the forces which
caused the acceleration. By analogy, we expect the Γ terms to represent
a counterpart to the force. When a force is derived from a potential,
it appears as a spatial derivative (with a minus sign) of the potential.
Denoting a potential by ϕ, we have:

$$f_x = -\frac{d\phi}{dx}$$

where f_x is the x component of $\vec{f}$. (Recall the connection between equations 6.5 and 6.3, of the gravitational potential and the gravitational field.)
In equation 7.4, the Γ terms are derivatives of the components of the metric tensor with respect to the coordinates, and they correspond to the force
terms in the Newtonian equation. The analogy is not complete, but we can
conclude that the components of the metric tensor play a role similar to
that of a potential from which a force can be derived. Hence these components are called "the gravitational potentials" and from them the general
gravitational effects are derived.

The components of the metric tensor play two roles: They characterize
the curvature of space, and they serve as potentials for the general gravitational effects. This double role of the metric tensor reflects the intercon-
nection between space curvature and the existence of gravitational effects.

In the correspondence between the relativistic mechanics and the Newtonian one, we may find a more explicit connection between the potentials
of the GTR and the Newtonian gravitational potential. Let us compare
intervals between two systems: One is an inertial system, and the other
is a system at rest in a gravitational field. Let us start our observations

when both systems coincide momentarily. Technically, this can be done by two parallel elevators, one of them held at rest against the gravitational attraction, and the other is released to fall freely. The free-falling one is an inertial system. At the moment of the release, both are at rest, and the world lines of the observers in the two elevators consist of time terms only. When the two systems coincide, both observers measure equal intervals and we have $dS^2 = dS_o^2$. This is equivalent to:

$$\sqrt{A_{44_o}}\, dt_o = \sqrt{A_{44}}\, dt \qquad (7.7)$$

where A_{44} is the component of the metric tensor which is the coefficient of the time differential in the interval, and the lower mark $_o$ denotes quantities measured in the inertial system. In the inertial system, all the diagonal components of the metric tensor are equal in their absolute values to unity (all the other components vanish), hence $A_{44_o} = 1$. We obtain from equation 7.7: $dt_o = \sqrt{A_{44}}\, dt$. This last term reminds us of the time transformation in the STR. (There we had : $dt_o = \sqrt{1 - v^2/c^2}\, dt$.) We expect that when dealing with a weak gravitational field, the GTR corresponds to the STR. Since the non-inertial system is such because of gravity, we expect that instead of v^2/c^2 which appears in the STR transformation, there will appear a term which expresses gravitation. Intuitively we can use the Newtonian relation between acceleration, velocity and the length of the path:

$$v^2 = 2 \cdot a \cdot l \qquad (7.8)$$

where a is the acceleration. When the acceleration is caused by gravitation, we have the connection (equation 6.7): $\Delta\phi = -a \cdot l$. Using this equation we insert equation 7.8 into equation 7.7 to find:

$$dt_o = dt\sqrt{A_{44}} = dt\sqrt{1 + \frac{2\Delta\phi}{c^2}}. \qquad (7.9)$$

Taking $\Delta\phi$ as the potential difference between the point considered and infinity (where $\phi = 0$), we write:

$$A_{44} = 1 + \frac{2\phi}{c^2}. \tag{7.10}$$

For a more general case, in which we have to consider relative velocity between the systems in addition to gravity, we write:

$$A_{44} = 1 + \frac{2\phi}{c^2} - \frac{v^2}{c^2}. \tag{7.11}$$

We can compare equation 7.10 with the parallel expression obtained for the rotating disc. The inertial force acting on an object rotating with the disc is a centrifugal force whose magnitude per unit mass is $\omega^2 r$. We may present this force as derived from a potential. In this case, it should be equal to the (minus) derivative of the potential with respect to the coordinate r. To obtain the centrifugal force from a potential, the potential ϕ_{cen} should be:

$$\phi_{\text{cen}} = -\frac{1}{2}r^2\omega^2. \tag{7.12}$$

Substituting into equation 7.10, we obtain:

$$A_{44} = 1 + \frac{2\phi}{c^2} = 1 - \frac{\omega^2 r^2}{c^2}$$

which corresponds to the result we obtained at the end of Chapter 5. When ω approaches zero, the centrifugal potential approaches zero too, and $A_{44} \to 1$, as in the Newtonian mechanics.

7.5.   Coordinate Time

A more consistent method for comparison of time intervals can be attained by the comparison of "proper time" and "coordinate time." As mentioned

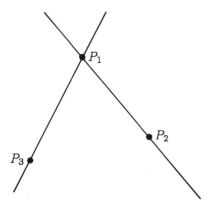

FIGURE 7.3.   Two geodesics intersect at the point P_1.

above, a proper time is the one measured by a clock located in the rest system of the observer, which means a clock carried with the observer along its world line. The proper time does not depend on the coordinate system chosen to describe the space-time, but we cannot compare proper times if they are measured by clocks carried along different geodesics. A geodesic is unique in the sense that it is completely defined by the point through which it passes and the object velocity at that point. (In a space-time diagram the velocity is characterized by the inclination of the line at the point.) Hence, if a free object moves along a trajectory which passes through the points P_1, P_2 and another free object moves along a trajectory which passes through the points P_1, P_3, then they are moving along different geodesics (with different velocities), and their proper times cannot be compared. In order to compare time intervals between the points P_2, P_3, we have to use coordinate time, which is the time coordinate of this system. The coordinate time is not an invariant, but its product by the corresponding component of the metric tensor is an invariant. This product can be integrated along the trajectory in the coordinate system.

The coordinate time is related to the proper time by:

$$d\tau = dt\sqrt{A_{44}}. \tag{7.13}$$

This equation exists at each point in space-time. Thus, the comparison of time intervals along different geodesics should be done through this relation. If we want to compare time intervals between the points P_2, P_3, we write:

$$\frac{d\tau_2}{d\tau_3} = \frac{dt\sqrt{(A_{44})_2}}{dt\sqrt{(A_{44})_3}} = \sqrt{\frac{(A_{44})_2}{(A_{44})_3}}. \tag{7.14}$$

In equation 7.14 $(A_{44})_2$ is a component of the metric tensor of a system which moves with an object whose geodesic passes through P_2, and the same is true for $(A_{44})_3$. The comparison between the proper time of these two objects is obtained by the ratio of the components of the metric tensors of the systems which move with these objects.

According to this equation, we can compare time intervals between a potential well and a point of zero potential. At the point where $\phi = 0$, $A_{44} = 1$. In a potential well we have: $A_{44} = 1 + (2\phi/c^2)$. Thus we obtain:

$$dt_o = dt\sqrt{1 + \frac{2\phi}{c^2}}.$$

When $(2\phi/c^2) \ll 1$, the right wing can be calculated by the binomial expansion:

$$(1 + x)^{\frac{1}{2}} = 1 + \frac{1}{2}x - \frac{1}{8}x^2 + \cdots$$

where $x = 2\phi/c^2$. Ignoring higher powers of x in the expansion, we obtain:

$$dt_o = dt \left(1 + \frac{\phi}{c^2} \right)$$

which is equal to what we have already obtained in equation 6.11.

7.6.   Geodesics on a Rotating Disc

Let us examine some interesting situations arising from the properties of the geodesics. Suppose that a frictionless object is lying on a rotating disc which rotates with a constant angular velocity ω. Because there is no friction between the object and the disc, it will not participate in the disc motion, and no centrifugal force acts on it. If this object is at rest with respect to the external system, it will remain so, and an observer who moves with the disc (the disc system is his rest system) will observe that the object is performing a motion along a circular orbit with angular velocity $-\omega$. For both systems, the object moves along geodesics because it is a free object. For an external observer, the situation is simple. The object is free and it stays at rest. The disc observer observes that the object moves in a circular orbit, which is a curved trajectory. Since he already knows that his system is a curved space (if he has read Chapter 5), he relates the character of this trajectory to the curvature of space. For him, the free object is moving along a geodesic which follows the curvature of space on the disc. Thus, relating the curved geodesic to the curvature of space yields a simple explanation to the phenomenon. The curved space is considered as a system of constraints which impose on the free object a curved trajectory.

If, instead of dealing with constraints, the disc observer prefers to interpret his observations by the action of forces acting on the object, the considerations will be complicated. Objects which are at rest in the disc system feel the action of a centrifugal force directed outwards. The motion of the free object (relative to the disc) with the angular velocity $-\omega$ implies that another force acts, opposing the force which usually acts on all objects in the disc system. An object moves in a circular orbit only if there exists a centripetal force (directed towards the center) which forces it to do so. The fact that the free object does not accelerate outwards (as do all the objects in the disc system), but, on the contrary, it accelerates inwards along the circular orbit can be explained by the existence of a force which fulfills two roles: first, it cancels the outward acceleration, and second, it imparts to the object an inward acceleration. The magnitude of the force per unit mass for each one of the two roles is $-\omega^2 r$. Hence this (fictitious) force per unit mass should be $2 \cdot (-\omega^2 r)$. This force is known as the Coriolis' force, (it is also an inertial force), and it is observed by observers at rest in rotating systems. We observe this force in our rotating system of the earth. The acceleration caused by this force is $2 \cdot \omega \cdot v$, where v is the linear velocity of the object relative to the rotating system. This force is perpendicular both to the linear velocity and to the vector of the angular velocity of the system. The direction of the angular velocity is along the axis of rotation.

Hence, the direction of the Coriolis' force is perpendicular to the object motion and in a plane orthogonal to the axis of rotation. This force will cause a deflection of the trajectory of the object with respect to its original direction. If the frictionless object entered the rotating system with some constant velocity (relative to the external system), it will continue with the same velocity. The disc observer will observe that the object performs a curved trajectory relative to the disc.

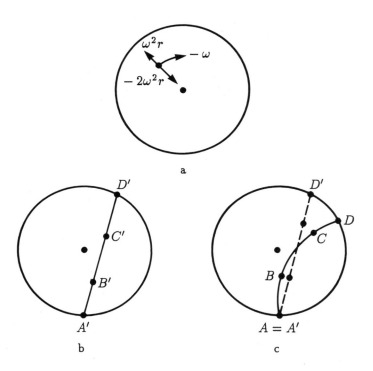

FIGURE 7.4.   A disc rotates counter-clockwise with angular velocity ω.

a.   A frictionless object lies at rest with respect to the external system. In the disc system the object is observed to rotate clockwise with angular velocity $-\omega$.

b.   An object moves in a straight line with respect to the external system and passes through the points A', B', C', D'.

c.   The points A, B, C, D are marked on the disc and rotate with it. They coincide with the points A', B', C', D', whenever the object has passed through each one of them. For the disc observer, they form a curved trajectory.

In Figure 7.4 we display a disc which rotates with angular velocity ω in a counter-clockwise direction. In Figure 7.4a, we see the free object which is at rest relative to the external system and the forces acting on it according to the disc observer interpretation. In Figure 7.4b we see the object which enters the disc system with constant velocity. Its trajectory passes along the points A', B', C', D', (laid along a straight line in the external system), but the object passes through these points at different times. The points A, B, C, D appearing in Figure 7.4c are marked on the disc and rotate with it. They coincide with A', B', C', D' respectively, when the object passes through each one of them. For the disc observer, the object's trajectory is described by the points A, B, C, D, and they do not describe a straight line. A', B', C', D' are on a straight line but at different times. Hence the disc observer argues that according to his observations the object moved along the line marked by A, B, C, D (Figure 7.4c). For him the velocity with which the object entered the disc is given by the vector addition of its linear velocity with respect to the external system and r/ω, which is the vector of the linear velocity of the disc at that point. The force interpretation for the curvature of the trajectory is again a Coriolis force. In the rotating system we see that this force imparts to each object an acceleration directed to the left of the motion, if the system rotates clockwise. Rotation of the system counter-clockwise causes an acceleration directed to the right of the direction of the object's motion. This effect can be observed in the rotating system of the earth, but due to its small magnitude (the angular velocity of the earth is: $\omega = 7.2722 \times 10^{-5}$ rad/sec), its influence is important only for the calculation of the trajectories of ballistic shells and missiles and for the motion of huge masses of air in the atmosphere over large distances. In the northern hemisphere, the rotation of the earth is counter-clockwise, and hence moving objects are deflected to the right from their trajectories. In 1926 Einstein wrote an article[5] entitled "The reason for the creation of curves in the paths of rivers." In this article Einstein discusses the phenomenon, which was known to geographers for long time, that the rivers in the northern hemisphere erode their right banks more than they do it to their left banks. The reason is the Coriolis' force, which together with other factors cause this phenomenon.

7.7.   Geodesic Triangle

The object we discussed in the preceding section behaves as a free particle because it is not acted upon by any external force. Hence its trajectories are geodesic, and the form of these geodesics reflects the space curvature of the rotating disc. A figure similar to Figure 7.4c can be drawn for numerous objects entering the disc from different directions. Thus we can draw a set of geodesics, and find the relations between them. In Figure 7.5, three such geodesics are drawn. The geodesics form a triangle between the points of their intersections (a geodesic triangle). For a disc observer, the geodesics are the "straight lines," and for him this geodesic triangle is the natural one. If we sum the angles of the triangle, we find that this sum is smaller than π (this can be verified by comparing the triangle in the figure with a rectilinear triangle, with the same vertices). Recall equation 4.5, where we had for the angles of a triangle constructed by geodesics:

$$\alpha + \beta + \gamma = \pi + Q \cdot A$$

where A is the triangle area, and Q is the surface curvature. Q is negative here and this again shows that the rotating system behaves as a space with negative curvature.

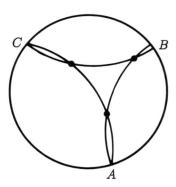

FIGURE 7.5.   Geodesic triangle. Three objects entered the rotating disc system at the points A, B, C. The intersections of their trajectories form a geodesic triangle whose sum of angles is smaller than π.

Let us follow the line of thought which led us to the present point.

We began with the generalization of the principle of relativity to include observations made by accelerated observers in the covariant picture of nature. The equivalence of gravitational phenomena and observations made in accelerated systems allowed us to include both types of phenomena in one class, which we called 'general gravitational phenomena.' All the phenomena in this class can be treated on the same footing. Since general gravitational effects affect all objects in the same way regardless of their mass, the influence of these effects can be represented by a geometric description. The general gravitational effects are translated into a set of constraints imposed on space-time, thus creating a curved space-time in which all physical events take place.

This curvature reflects the fact that the motion of any object is influenced by its interactions with other masses in the universe; the static interaction is known as the Newtonian gravity, and the dynamic one appears as the inertial properties of material objects. We may treat the curvature as a field induced on space by the presence of matter in the universe. This field serves as the carrier of the interactions of the material object with all the other matter in the universe. The curvature includes all the properties of space and leads free objects to move along geodesic trajectories. These properties of space are characterized by the metric tensor. Hence, the components of the metric tensor are the variables of the GTR.

7.8.   Einstein's Equations

The missing link for the completion of the picture is to show how the components of the metric tensor are created by the distribution of matter in space-time. Einstein's equations are the equations which connect the components of the metric tensor to the distribution of matter in space-time.[6] Formally, Einstein's equations are equations of tensors of rank 2 (represented by 4×4 matrices). On the left wing there appears a tensor called Einstein's tensor, denoted by **G**, whose components include the components of the metric tensor and their derivatives. On the right wing of the equation we have the energy-momentum tensor, denoted by **T**, multiplied by a constant. The components of **T** include the mass-energy density, the momentum density, and the components of the shear tensor.

A tensor equation should be valid for each pair of corresponding components on both side of the equation. A four-dimensional tensor includes

$4 \times 4 = 16$ components. Hence such a tensor equation is actually a set of sixteen equations for sixteen variables. Because of the symmetry of the metric tensor, **G** is symmetric too ($G_{ij} = G_{ji}$). Because of this symmetry we are left with ten independent equations. Usually, the higher symmetry of the specific cases which we deal with creates additional symmetries, and these symmetries reduce further the number of independent variables and equations.

What requirements can we put on these equations?

According to the correspondence principle we adopted, we demand that for low velocities and weak gravitational fields the GTR equations will reduce to the known Newtonian equation of gravitation. By this correspondence we can determine the coefficients appearing in the equations. Hence, although Einstein's equations are more general than Newton's, they will have a similar form. Further requirements are that the equations do not depend upon a specific coordinate system and that they yield the general conservation laws of physics.

The Newtonian equation of gravitation is derived from the relation between the gravitational force field and the potential. The force field equals the (negative) derivative of the potential with respect to the coordinates. (This can be verified by differentiating equation 6.5 with respect to r, to obtain equation 6.3). The rate of change of the field with respect to the coordinates is given by its derivative with respect to the coordinates, and it depends upon the distribution of the sources of the field (which is the mass density distribution). This means that the first derivatives of the field with respect to the coordinates are proportional to the distribution of the mass density. As the field itself equals (negative) first derivatives of the potential, the derivatives of the field equals the second derivatives of the potential (with a minus sign), and this should be equal to the distribution of the mass density. Hence, the Newtonian equation, known as Poisson's equation, includes on its left wing second derivatives of the potential with respect to the coordinates (with a negative sign), and on its right wing it includes the mass density distribution. This is formulated by:

$$\nabla^2 \phi = -4\pi G \rho \qquad (7.15)$$

where ϕ is the potential, ρ is the mass density (in Newton's theory, only rest mass is considered), G is the constant of gravitation, and ∇ is an operator which represents differentiation with respect to the coordinates. (The symbol ∇^2, denoted Laplacian, means that this operator is applied twice). The solution of equation 7.15 determines the gravitational potential at each point in space, as induced by the mass density which appears on the right wing of equation 7.15.

The general form of Einstein's equations should be similar to equation 7.15, including time derivatives in addition to the spatial derivatives. The equations are formulated as:

$$G_{ij} = JT_{ij} \tag{7.16}$$

which means that each component of **G** is equal to the corresponding component of **T** times the constant J (J is equal to $-4\pi G/c^2$). The solution of equation 7.16 determines the components of the metric tensor at each point in space-time, as induced by the energy-momentum distribution which appears in T_{ij}.

From the considerations given in the preceding Chapter, we know that the gravitational potential appears in the component A_{44} of the metric tensor ($A_{44} = 1 + 2\phi/c^2$). We require that for low velocities and weak gravitational fields (the Newtonian approximation) the equation between the component 44 of **G** and **T** ($G_{44} = JT_{44}$) will be parallel to Newton's equation. Hence, the tensor **G** should include second derivatives (and not of a higher order) of the components of the metric tensor. The same considerations lead to the idea that T_{44} should include the matter density ρ.

What are the other components of **T**?

The mass density included in T_{44} is the mass density measured in the Lorentzian system of the observer. It includes the rest mass of matter plus all kinds of energies which contribute to the gravitational effects (except for the gravitational energy). The other components in the fourth row (T_{4i}; $i = 1, 2, 3$), include the momentum density. Thus the fourth row is actually the four-dimensional momentum vector (mentioned in Chapter 4), as measured in the observer system. When the mass density is at rest, the 1, 2, 3 components of this vector vanish. Due to the symmetry, the fourth column of this tensor equals the fourth row.

The other nine components of T_{ij} ($i = 1, 2, 3$; $j = 1, 2, 3$) are the components of the shear tensor as described in Figure 2.2 (Chapter 2), as measured in the observer system. The normal stress (orthogonal to the cube face in Figure 2.2) is the pressure. Hence, the diagonal components of this tensor represent the pressure in the system.

As an example, let us present the energy-momentum tensor of a perfect fluid at rest. (A perfect fluid is one in which, except for pressure which emerges from elastic interactions between its particles, no internal forces such as viscosity exist). Because of the perfectness of the fluid, the non-diagonal components of the shear tensor vanish, and we have only the pressure terms along the diagonal. As the fluid is at rest, the components of momentum density vanish, and only the fourth component of this vector survives—the mass-energy density. The diagonal tensor **T** is given by:

$$\mathbf{T} = \begin{pmatrix} P & 0 & 0 & 0 \\ 0 & P & 0 & 0 \\ 0 & 0 & P & 0 \\ 0 & 0 & 0 & \rho c^2 \end{pmatrix}. \qquad (7.17)$$

Recall the four-vector of the velocity ($U = \gamma\binom{\vec{v}}{c}$, and in the present case $\vec{v} = 0$ and $\gamma = 1$). Hence, equation 7.17 can be written in a more general and compact form:

$$T_{ij} = (\rho + P/c^2) \cdot U_i \cdot U_j - P \cdot A_{ij} \qquad (7.18)$$

where U_i is the ith component of the velocity four-vector.

7.9.   Non-Linearity of Einstein's Equations

There is a major difference between Einstein's equations (the field equations of the gravitational field) and Maxwell's equations (the field equations of the electromagnetic field). Einstein's equations are non-linear while Maxwell's equations are linear.

One reason for this non-linearity in the equations is the fact that they include the term Γ, and we have already seen that this term is non-linear in the components of the metric tensor. A second reason for the non-linearity is the way in which we calculate derivatives in a curved space. We required that the equations correspond to the Newtonian equation, and since that equation includes second derivatives with respect to the coordinates, Einstein's equations should also include second derivatives in a linear form. We have seen in section 3 that when a second differentiation is performed, there must appear products of the first derivatives of the metric tensor (equation 7.6). These products add to the non-linearity of the equations. In a flat space, $\Gamma = 0$, and all the non-linear terms vanish. In a space which has a weak curvature (the gravitational fields are weak), the Γ terms are very small, and their products can be ignored. With this we obtain the "weak field approximation" which yields solutions which approximate the Newtonian solutions.

Thus we understand how the non-linearity appears on the left wing of the field equations ($\mathbf{G}$ includes the components of the metric tensor and their derivatives). In order to understand how the non-linearity appears on the right wing of the equations, let us examine equation 7.16. At first glance the equation's left wing seems to include the dependent variables and their derivatives (the components of the metric tensor and their derivatives), and the right wing includes the sources for the "behaviour" of these variables, where the sources are the momentum and the energies (rest mass + energies) of the matter in the universe. Such an impression fits well our comprehension of the parallel Newtonian equations. But in the relativistic equation the situation is different.

In classical physics (Newtonian mechanics and electromagnetism), we work in a three-dimensional space, and time is a parameter for measuring the evolution. Take, for example, Newton's equation of motion, where we obtain the acceleration of an object as a function of its mass and the forces acting on it. Usually, the forces depend upon the location of the object. Hence, the acceleration calculated by solving the equation of motion tells us how its location is changed with time and, as a result of the location change, the forces acting on it will be different. But, at a given moment, there is a definite situation which determines the acceleration at that moment. The change created as a result of this acceleration is moving to a new situation at a later moment. For the new situation (at the later moment) we write a new equation, and we solve it for a new acceleration which again changes the location, and so on. At each moment there exists a given configuration of masses and forces, which can be inserted on the right wing of the equation of motion, and the left wing yields the acceleration. By solving a set of equations for consecutive configurations, we obtain the time evolution of the location of the object. Such an equation is linear, as any change in the acting factors (right wing of the equation) creates a proportional change in the consequence (left wing of the equation).

In relativistic mechanics the situation is different. We work in a four-dimensional space which is the space where the events occur. The equations which describe a state of an object include all the events in which the object had participated in the past, and all those in which it will participate in the future. Einstein's equations are equations for the components of the metric tensor, and their solution yields the state of these components over the whole four-dimensional space. These components express the curvature of space, which means the constraints imposed on matter in this space, and thus they determine the behaviour of matter, which appears on the right wing of the equations as a source. As a result, the dependence between the two wings is a two-way dependence, and hence the equation is non-linear. We can imagine a state in which a certain change in the source (the matter density) creates a very big change in the components of the metric tensor. If the gravitational field is strong, the components of the metric tensor will be very different from their values in the STR, and the changes in the state of matter may be fast. In such a case, the two-way dependence between the two wings of the equation is strong, and the non-linearity of the equation is strong too. When the gravitational fields are weak, the two-way dependence is weak too, and usually it can be ignored. The result appears in the behaviour of the sources. **T**, which represents the sources of the gravitational effects, includes all the masses and energies, including kinetic energy and energies connected with angular momentum (except for gravitational energy). If the motion of objects which serve as sources for the field is influenced by the field, their energies may change, and thus change the sources. Hence, a condition for the linearity of the equations is that the sources do not change, which means that the given configuration of locations and motions does not change.

Another difference between the classical theories and the GTR is that in the electromagnetic theory, we have Maxwell's equations (the field equations for the electromagnetic fields). These equations are linear. In addition, there is the force equation (Lorents force), which describes the interaction between the charges. It is similar to the Newtonian gravitation, where we have a linear equation for the field (equation 7.15) and another equation for the force between the masses (equation 6.1). In both these theories, the equations of motion are derived from the equation of the force.

In the GTR, Einstein's equations for the components of the metric tensor are the equations for the gravitational field. Their solution yields the components of the metric tensor. We have already seen that from these components we obtain the geodesic trajectories, which actually are the equations of motion of objects due to the general gravitational effects. Hence, in Einstein's theory, we have no additional equation of force. All the information needed for obtaining the equations of motion is included in the field equations.

We can understand now what the two conditions are for the correspondence between the GTR and Newtonian theory. One condition is low velocities. This condition asserts that effects connected with the STR (terms including v^2/c^2) can be ignored. The second condition is a weak gravitational field. This demand asserts that the non-linear part of the equations is very small and can be ignored relative to the linear part, and then gravity can be dealt with like a linear theory. The gravitational theory of Newton is linear and deals with cases which can be approximated as static cases. The gravitational theory of Einstein is non-linear and deals with dynamic situations, and it corresponds to the Newtonian theory when it deals with low velocities and weak gravitational fields.

Comments and References

1. Coxeter, H. S. M. *Introduction to Geometry.* John Wiley and Sons, Inc., New York, 1961.

2. Feynman, R. P. *Lectures on Physics.* Addison-Wesley, Reading, Massachusetts, 1963, ch. 19.

3. We use the term Γ in a loose form. In professional textbooks the authors strictly keep the difference between different forms of Christoffel symbols. Actually the sign Γ is used for the Christoffel symbol of the second kind.

4. A short movie entitled "Systems of Reference," produced by the PSSC group, displays this behaviour in a very convincing way.

5. Einstein, A. *Ideas and Opinions.* Crown Publishers Inc., New York, 1954.

6.   From the STR we know that mass and energy are equivalent, and their interrelation is given by $E = mc^2$. "Pure" energy is influenced by gravity, it has inertial properties, and it is a source for gravitational interactions. We shall use further the concept of mass or matter to denote mass-energy, even if it is not stated explicitly.

CHAPTER EIGHT

Schwarzschild's Solution

Einstein's equations described in the preceding chapter are the equations whose solution yields the components of the metric tensor which characterizes the properties of space and the behaviour of matter in it. These equations are non-linear partial differential equations of the second order. There is no general solution for these equations. In a regular differential equation, there is one independent variable, and the other variables are all dependent upon this variable. In most of the cases, a regular differential equation can be solved by integration. A partial differential equation includes several independent variables, and integrating such an equation is hardly possible. In some simple cases, a partial differential equation can be decomposed into several regular differential equations, where each one depends upon one independent variable. Usually, such cases can be solved analytically. In Einstein's equations we have four independent variables—the four coordinates (x, y, z, ct)—and the components of the metric tensor which are the dependent variables.

As there is no general solution to Einstein's equations, they are treated by investigating specific cases, in which we have some advance feeling for the form of the solution. Such cases occur where the solution is conditioned by some symmetry, some preliminary assumption about the matter distribution, etc.

We present here a solution found by Schwarzschild in 1916 for an isolated object. This solution assumes two preliminary assumptions: the isotropy of space and the existence of a static solution. The solution deals with a finite amount of mass (a single object); hence, it is convenient to choose the center of mass as the origin of the coordinate system. Due to the isotropic assumption, there should be spherical symmetry around the center; a convenient choice will be to work with a spherical coordinate system (r, θ, φ). The meaning of the spherical symmetry is that there is no dependence on the angular coordinates (θ, φ). The assumption of a static solution excludes dependence on time. Hence, we are left with only one independent variable, r, which is the distance from the center of the mass. (This treatment is an example for a case in which certain preliminary assumptions enable us to find a solution for this certain case.)

In a flat space the line interval of a spherical system is given by:

$$dS^2 = c^2 dt^2 - dr^2 - r^2(d\theta^2 + \sin^2\theta d\varphi^2). \tag{8.1}$$

When there is no dependence on the angular coordinate we may write: $(d\theta^2 + \sin^2\theta d\varphi^2) = d\Omega^2$ and later on we shall use this term. In a non-flat space (with the spherical symmetry assumed), a general form for the interval is:

$$dS^2 = Hc^2 dt^2 - (Bdr^2 + Dr^2 d\Omega^2) \tag{8.2}$$

where the diagonal component of the metric tensor are: $H = A_{44}$; $-B = A_{11}$; $-Dr^2 = A_{22}$; $-Dr^2 \sin^2\theta = A_{33}$, and H, B, D are functions of r only. The interval includes no mixed products of the coordinates, and the off-diagonal components of the metric tensor vanish. In equation 8.2 we have three variables to be found, but we still have the freedom to adjust r. For each r we choose with a given D, we can choose a new $r : r' = r \cdot \sqrt{D}$, and by this substitution we are left with two dependent functions only: H, B. According to the preliminary assumption, they depend on r only: $B = B(r)$, $D = D(r)$.

Schwarzschild's solution treats separately two regions of space: the exterior solution deals with the empty space around the mass which resides around the origin, and the interior solution deals with space properties inside the matter distribution. We demand that the two solutions will be continuations of each other, and thus they should coincide at the boundary between the two regions, which is the mass boundary.

8.1.   The Exterior Schwarzschild Solution

We begin with the exterior solution which may serve as a good approximation for a single mass which is far away from all other masses. This approximation suits well our solar system. Most of the matter in the solar system is concentrated in the sun (more than 0.998 of the total mass of the solar system). The planets, whose masses are negligible compared to the sun, may be considered as test particles in this system. The magnitude of the solar system, up to the most distant (known) planet is a few light-hours, while the distance to the nearest star is about four light-years. Hence, in the solar system we can ignore the static gravitational potential of other stars and deal only with potential of the sun.

As we deal with the empty space around the central object, the components of th e energy-momentum tensor, T_{ij}, vanish for this region, and Einstein's equation reads:

$$G_{ij} = 0. \tag{8.3}$$

Because of the symmetry, all the off-diagonal components of **G** vanish, and we are left with diagonal components only, G_{ii}, $i = 1, 2, 3, 4$. The two components connected with the angular coordinates are proportional to each other ($G_{22} \propto G_{33}$), and we have three equations: $G_{11} = G_{22} = G_{44} = 0$.

When we calculate these equations, we find:

$$-2\frac{B}{H}G_{44} = \frac{H''}{H} - \frac{1}{2}\left(\frac{H'}{H}\right)^2 - \frac{1}{2}\frac{H'}{H}\frac{B'}{B} + 2\frac{H'}{Hr} = 0 \tag{8.4a}$$

$$2G_{11} = \frac{H''}{H} - \frac{1}{2}\left(\frac{H'}{H}\right)^2 - \frac{1}{2}\frac{H'}{H}\frac{B'}{B} - \frac{2B'}{Br} = 0 \tag{8.4b}$$

$$BG_{22} = 1 + \frac{r}{2}\left(\frac{B'}{B} - \frac{H'}{H}\right) - B = 0 \tag{8.4c}$$

where $H' = dH/dr$; $H'' = d^2H/dr^2$.

As similar terms appear in all the equations, we can obtain simpler ones by arithmetic manipulation between the equations. Subtracting the second equation from the first one, and multiplying by $r/2$ we find: $H'/H + B'/B = 0$. Using this relation to substitute B'/B into G_{44}, we obtain an equation for H:

$$\frac{H''}{H} + \frac{2H'}{Hr} = 0. \tag{8.4d}$$

This equation is a regular differential equation for the variable H. Its solution is:

$$H = I - \frac{2m}{r} \tag{8.5}$$

where I and m are constants of integration whose value is to be determined later when we compare the relevant components of the metric tensor to the corresponding Newtonian terms. (The solution can be verified by differentiating H once and twice with respect to r, and inserting H', H'' into equation 8.4d.) The expression we obtained earlier ($H'/H + B'/B = 0$) is a simple equation whose solution is: $H \cdot B = $ constant. The value of this constant is found by the value it should have at infinity. When $r \to \infty$, we are at an infinite distance from the mass, where its potential approaches zero, and the value of H, B should approach their value in empty space, which is unity. Hence $H \cdot B = 1$. Further, $2m/r$ approaches zero at infinity. From this considerations we find:

$$H = 1 - \frac{2m}{r} \tag{8.6a}$$

$$B = \frac{1}{H} = \frac{1}{1 - \frac{2m}{r}}. \tag{8.6b}$$

By inserting into equation 8.4c the expressions obtained for H, B we get an identity. We find that from the three equations we have, only two are really independent, and the third one adds no information. Inserting the expressions of H, B into the interval yields:

$$dS^2 = \left(1 - \frac{2m}{r}\right)c^2 dt^2 - \frac{dr^2}{1 - \frac{2m}{r}} - r^2 d\Omega^2 \qquad (8.7)$$

and the metric tensor is:

$$\mathbf{A} = \begin{pmatrix} \frac{-1}{1 - \frac{2m}{r}} & 0 & 0 & 0 \\ 0 & -r^2 & 0 & 0 \\ 0 & 0 & -r^2 \sin^2 \theta & 0 \\ 0 & 0 & 0 & 1 - \frac{2m}{r} \end{pmatrix} \qquad (8.8)$$

At a large distance from the mass, $r \to \infty$, $2m/r \to 0$, $B \to 1$, $H \to 1$, and the metric tensor reduces to its form in a flat space. The non-flat contribution in H, B falls to zero as $1/r$. These terms show by how much the components of the metric tensor differ from their values in a flat space, and this gives a measure to the source which creates the space curvature. We found in the preceding chapter that the Newtonian potential is included in the metric tensor by: $A_{44} = H = 1 + 2\phi/c^2$. Comparing this expression to 8.6a shows that the constant m is proportional to the mass, and its value is:

$$m = \frac{GM}{c^2} \qquad (8.8a)$$

where M is the total mass of the object at the origin, and G is the gravitational constant.

8.2. Application of Schwarzschild Solution to the Solar System

As mentioned above, we treat the planets in the solar system as test particles with negligible masses, particles which move in a spherical symmetric static gravitational field at large distances from the central mass. Since the only force acting on the planets is the gravitational force, they should move along geodesics in a curved space whose curvature is characterized by the components of the metric tensor. Let us calculate the geodesics equations in this space and compare them to the parallel Newtonian equations (obtained from the Newtonian equation of motion for the trajectories of the planets). By this comparison we can examine the similarity and the difference between the GTR and the Newtonian theory, and we can determine the value of the constants which appear in the relativistic equations.

Because the gravitational force is central (the force is always directed towards a fixed point, to the center of gravity), each planet moves in a fixed plane around the sun. We can choose the coordinate system so that the plane of motion of the planet is at the equatorial plane of the system. By this choice we have:

$$d\theta = 0; \qquad \theta = \pi/2; \qquad \sin\theta = 1$$

and the motion in this plane is characterized by r (the distance of the planet from the center), and by φ, (the angle between the line connecting the planet to the center) and the zero line chosen arbitrarily (polar coordinate system in a plane, Chapter 3). In the Newtonian calculation we have two equations which determine the time dependence of r and φ. The time can be eliminated between the two equations, and we can obtain an equation called the 'orbit equation' in which r is given as a function of φ (the distance of the planet from the sun at any given angle). The Newtonian solution to the orbit equation is:

$$r = \frac{r_o}{1 + \varepsilon \cos\varphi} \tag{8.9}$$

which is an equation of an ellipse or a hyperbola, r_o is a constant, and ε is the eccentricity of the line. The character of the line is determined by ε. The expression 8.9 is a general expression which describes the orbit of an object in a plane. All the solutions of this form are called "sections of the cone" because they can be obtained by an intersection of a cone with a plane. When $\varepsilon = 0$, we have $r = r_o$, which is the equation of a circle.

The circle is obtained from the intersection of a cone with a plane which is parallel to the cone's base. If the plane is inclined by some angle to the

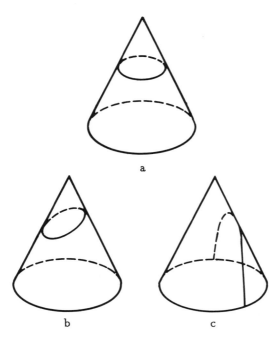

FIGURE 8.1.   The sections of the cone. Formed by the intersection of a plane and a cone.
    a.   The intersecting plane is parallel to the cone base.
    b.   The intersecting plane is inclined to the cone base. The intersection line is an ellipse.
        $0 < \varepsilon < 1$.
    c.   The intersecting plane is inclined to the cone base by an inclination which is steeper than
        the generating line of the cone. The intersection line is a hyperbola, $1 < \varepsilon$.

cone basis, the intersection of the cone with the plane forms an ellipse. This case is obtained for $0 < \varepsilon < 1$. Once the inclination is such that the plane is parallel to the generating line of the cone, the intersection will not be a closed line any more. This case is obtained when $\varepsilon = 1$, and the intersection line is a parabola. For further inclination of the plane, the intersection line will be a more opened line. Such a case is obtained with $\varepsilon > 1$, and the lines obtained are hyperbolae.

The geodesic equations are equations of orbits because they describe the orbit which an object will take at the given space, regardless of its mass. Hence it is convenient to compare the equations of the geodesics to the Newtonian equations of the orbit, which do not include the mass of the object. Hence, we divide the corresponding Newtonian equations by the object's mass and obtain all the variables in units per unit mass (specific variables). Thus we have specific energy which is energy per unit mass, specific momentum, specific angular momentum, and, instead of force, we obtain the acceleration imparted by the force. On examining the Newtonian equation for the orbits of the planets around the sun, we find that ε and r_o are related to the specific energy and the specific angular monentum of the object moving in the orbit:

$$\varepsilon = \sqrt{1 + \frac{2EL^2}{G^2M^2}}$$
$$r_o = \frac{L^2}{GM}$$

(8.9a)

where M is the solar mass, E is the specific energy, and L is the specific angular momentum. The total specific energy, which is the sum of the kinetic and potential energies, can be negative or positive. The potential energy which causes the attraction is negative, while the kinetic energy (connected to centrifugal acceleration, which repels the object outwards) is positive. When the total energy is negative, the absolute value of the potential energy is larger than the kinetic energy. The attraction overcomes the repulsive centrifugal forces, and the object is bound to the center of gravity. It will move in closed orbits (a circle or an ellipse) and will not leave the system. When the total energy is positive, the kinetic energy is larger than the absolute value of the potential energy. The attractive forces are weaker than the centrifugal forces, the orbit is opened (a parabola or a hyperbola), the influence of gravity will only cause a deflection of the original orbit, and, after the encounter with the central mass, the object will move away to infinity.

Mathematically, the description of the orbit is given by the value of ε. When the energy is negative, the expression under the root sign in equation 8.9a is smaller than unity, $\varepsilon < 1$, and the orbit is an ellipse. When the energy is positive, the expression under the root sign is larger than unity, $\varepsilon > 1$, and the orbit is a hyperbola. The case $E = 0$ is a limit case: The absolute value of the potential energy is equal to the kinetic energy, $\varepsilon = 1$, and the orbit is a parabola. When we calculate the Newtonian equation of

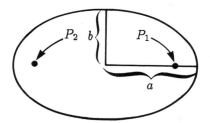

FIGURE 8.2.   An ellipse. P_1, P_2 are the ellipse's focuses.
  a is the semi major axis.
  b is the semi minor axis. ε, the eccentricity: $\varepsilon^2 = 1 - b^2/a^2$.

the orbit, we find r as a function of φ:

$$\left[\frac{d}{d\varphi}\left(\frac{1}{r}\right)\right]^2 = -\frac{1}{r^2} + \frac{2E}{L^2} + \frac{2GM}{L^2 r}. \tag{8.10}$$

The general conservation laws are the laws of conservation of the momentum, the energy and the angular momentum, and they must be valid in the theory of relativity. According to these laws, these three entities are constants of motion in isolated systems. The solar system is approximately an isolated system on which no external forces act. Hence, E and L are constants. Expression 8.9, which describes the orbits of objects in a field of a central mass, is the solution to equation 8.10. (This can be verified if we write: $1/r = (1 + \varepsilon \cos \varphi)/r_o$, differentiate it with respect to φ, and substitute r_o and ε according to equation 8.9a.)

When we calculate in parallel the equation of the geodesic orbit by the calculation of the variations of the interval given in equation 8.7, we obtain a differential equation which relates r and φ and which is parallel to equation 8.10. The relativistic equation obtained is:

$$\left[\frac{d}{d\varphi}\left(\frac{1}{r}\right)\right]^2 = -\frac{1}{r^2} + \frac{2E}{L^2} + \frac{2GM}{L^2 r} + \frac{2GM}{c^2 r^3}. \tag{8.11}$$

In this calculation E and L were obtained as constants of integration, and their meaning as specific energy and specific angular momentum are attributed to them in correspondence with the Newtonian equation 8.10. We find that equations 8.10 and 8.11 are almost identical except for an extra term in equation 8.11 $(2GM/c^2r^3)$. When we estimate the contribution of this term relative to the contributions of the other terms, we observe that in the denominator of this term appear c^2r^3. At large distances characteristic of the solar system, this is a very large quantity. Hence, the additional term is very small relative to the other terms, and its contribution may be considered as a relativistic correction to the classical solution. Using the Newtonian equation 8.10 is like using equation 8.11, if we ignore the last term. To find a measure for the magnitude of the correction, let us divide the last term by the term which precedes it in the equation. we write:

$$\left(\frac{2GM}{c^2r^3}\right) / \left(\frac{2GM}{L^2r}\right) = \frac{L^2}{c^2r^2}. \tag{8.12}$$

For an object moving in a circular orbit, the angular momentum L equals the product of the linear velocity and the radius of the orbit: $L = r \cdot v$. Hence, we obtain for the ratio of the fourth term to the third one (equation 8.12): $L^2/c^2r^2 = v^2/c^2$. For the earth, the linear velocity is 30 km/sec and the value of this ratio is 10^{-8}. Hence, the solutions of equations 8.10 and 8.11 will be very similar. As we shall see later, the correction required due to the contribution of the last term is very small, and its influence is significant only over a time scale of centuries. The situation will be different in systems in which the distances between the components of the system are much shorter. When r is small, the contribution of the last term will be larger, and it might have a significant influence. Later we shall see such an example.

8.3.   Tests for the GTR

Several conclusions drawn from Schwarzschild's solution may serve as a test for the GTR. Three of them were already known in the early epoch of the development of the theory of relativity: the precession of the perihelion of the planets, the bending of light trajectories in a gravitational field, and the red shift of light spectrum due to gravity. Recently, with the development of new equipment and new methods of measurement in astronomy, these effects can be measured for a wider range of phenomena and with higher precision. Let us describe these predictions in more detail:

a. Precession of the perihelion of the planets' orbits

The planets move around the sun in an elliptical orbit, and the sun is at one of the ellipse's focuses. The degree of ellipticity is given by the eccentricity, ε. The perihelion of an orbit is the point at which the planet is nearest to the sun. This point is located on the major axis of the ellipse, which passes through the two focuses of the ellipse. It has been found by observations for many centuries that the major axis of the orbits of the planets is not always aligned in the same direction, but rotates slowly around the sun, and the perihelion rotates with it. The rotation of the major axis is called precession. This phenomenon can be observed more clearly when the eccentricity is larger.

The difference between the Newtonian equation of the orbit and the relativistic one is in the last term of equation 8.11. Let us deal with the Newtonian equation as a first approximation, and study later what magnitude of correction is contributed by the additional term. The solution of the Newtonian equation (given in equation 8.9) describes an ellipse of the planet's orbit.[1] According to this solution, the direction of the major axis should remain constant. If we find that the precession of the perihelion is caused by the contribution of the additional term in the relativistic equation, then its existence is a verification of the GTR, which could predict this precession.

The solution of the relativistic equation 8.11 is:

$$r = \frac{r_o}{1 + \varepsilon \cos(\varphi - \Delta\varphi)} \qquad (8.12.a)$$

where the difference between the two solutions is in the addition of $-\Delta\varphi$ to the argument in the denominator, and this addition is time-dependent. This solution is also an equation of an ellipse, but as the argument of the cosine changes with time, the direction of the major axis changes too. The rate of the precession can be calculated. The planet reaches the perihelion each time the argument of the cosine equals some integer multiplied by 2π ($\varphi - \Delta\varphi = 2\pi \cdot n$, where n is some integer) or each time the planet completes a period of $2\pi + \Delta\varphi$. The planet completes an ellipse when it performs a complete cycle, which is a revolution of 2π. During this time, the perihelion has rotated by an angle of $\Delta\varphi$. The ratio $\Delta\varphi/\varphi$ represents the ratio of the rate of the rotation of the perihelion to the rate of rotation of the planet. When we insert the relativistic solution to the equation we obtain:

$$\frac{\Delta\varphi}{2\pi} = \frac{3GM}{c^2 a(1 - \varepsilon^2)} \tag{8.13}$$

where a is the semi major axis of the ellipse. The numerical value of $\Delta\varphi/\varphi$ is very small (its value is about 3×10^{-8}). Actually, the correction contributed by this term is significant and can be verified by observations only for the planet Mercury, and this is for two reasons: first, the small distance of Mercury from the sun (small a), and second, the large value of its eccentricty, which is $\varepsilon = 0.216$, while in the other planets their eccentricity is much smaller.

Actually there are several other factors which influences the precession: the fact that the sun is not an exact sphere, the influence of the other planets, etc. The precession which is caused by all the other factors can be calculated very exactly, and it was calculated in the 19th century by Leverrier. The value calculated was about 500 '' per century. (The notation '' denotes arc second which is 1/3600 of a degree. The notation ' denotes arc minute which is a 1/60 of a degree.) Observing Mercury, it was found that its precession is larger than the one calculated by $43''$ per century. When we insert the numerical values in equation 8.13 we find that $\Delta\varphi = 42.9''$ per century, a result which is in a nice agreement with what was expected.

b. Bending of the trajectory of a light ray in a gravitational field

This topic may be treated by calculating the geodesic line of a light ray in the sun's vicinity. The interval along light trajectory vanishes:

$$dS^2 = 0. \tag{8.14}$$

The geodesic calculated by the variations of this interval yields the equation of the trajectory of the light. The equation obtained is:

$$\left[\frac{d}{d\varphi}\left(\frac{1}{r}\right)\right]^2 = D^2 - \frac{1}{r^2} + \frac{2GM}{c^2 r^3} \tag{8.15}$$

where D is a constant of integration, and all other symbols have the same meaning as in equation 8.11. Comparing equations 8.15 with 8.11, we find that the two equations are quite similar if we write $D^2 = 2E/L^2$ (as L, E are constants), except that the term $2GM/L^2r$ does not appear in equation 8.15. GM/r is the term from which the static gravitational force is derived. This means that light is not influenced by the static gravitational force. The last term in equation 8.15 is the same as the last term in equation 8.11, where this term is the relativistic correction to the Newtonian equation. This term is very small relative to the other terms, and on a first approximation may be ignored to give the solution of equation 8.15:

$$r \cos \varphi = \frac{1}{D}. \tag{8.16}$$

This solution can be verified by differentiating the term $1/r = D \cdot \cos \varphi$ with respect to φ, and substituting the derivative into equation 8.15. Transforming from polar Coordinate system to Cartesian by substituting:

$$x = r \cos \varphi \tag{8.17}$$

we find:

$$x = \frac{1}{D} = x_o \tag{8.18}$$

where x_o is constant. The meaning of equation 8.18 is that the approximated solution yields the equation $x = x_o$ for light trajectory, which is the equation of a straight line. According to the Newtonian theory, light moves in space along straight lines. The additional term, which we ignored in this approximation, causes a deflection from the straight line trajectory. At large distances from the sun, $r \to \infty$, the trajectory will restore its straight line form, but with some inclination to the original straight line along which the sun was approached. We calculate the angle of deflection (which is the angle between the original straight line $x = x_o$) and the straight line which is an asymptote to the new trajectory, after the deflection (see Figure 8.3).

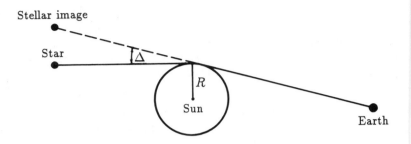

FIGURE 8.3.   Bending of a light ray due to gravity. A light ray emitted by a star is bent on passing near the solar edge. The observer on earth see the image of the star on the line connecting his eye and the bending region. The symbol Δ is the angle between the original trajectory and the asymptote to the trajectory after the bending.

Denoting the angle between the two lines by Δ, we find:

$$\Delta = \frac{4GM}{c^2 x_o} \tag{8.19}$$

where x_o is the orthogonal distance of the original trajectory from the solar center. Inserting numerical values into the solution, we choose for x_o the radius of the sun, which is the minimal value available. Actually we are dealing with a light ray which is just grazing the sun. We find:

$$\Delta = 1.75''. \tag{8.20}$$

Measuring such a deflection is very difficult in visible light. We have to observe star light when it passes close to the sun's edge, but normally, stars cannot be seen near the sun. The only possibility is to perform this observation during a complete solar eclipse. For this purpose the sun and its vicinity are photographed during the eclipse, and the picture is compared to the picture of the same part of the sky taken without the sun. The differences between the two pictures in the position of the stars are due to the presence of the sun in one of them. The distance to the fixed stars is very large relative to the distance of the sun from us, and the sun influences the trajectory of light coming from these stars when it moves near the sun. In Figure 8.3 the positions of the sun and a fixed star are drawn relative to our location.

This observation was first made by Eddington in 1919, and the prediction of the GTR was verified. Since then it has been repeated many times, and the prediction was always confirmed. To date, measurements with other kinds of electromagnetic radiation, such as radio emission from quasars, allow us to measure this effect without the need for an eclipse and with higher precision. All these measurements definitely confirm the deflection of the radiation according to prediction of the GTR[2].

The importance of this prediction is that the phenomenon of light deflection was not known before. The phenomenon of the precession of the perihelion, explained in the preceding section, has been known for long time, and the GTR supplied the lacking explanation for the phenomenon. The deflection of light trajectory is an original prediction of the GTR. An explanation for a known and unexplained phenomenon is important, but it is not such a strong test of the theory because people may always assume that the theory was created to explain the already known phenomenon. But the prediction of an unknown phenomenon, and its verification by observations, is a very strong test of a theory, and this is the importance of this test.

The deflection of light in a gravitational field may be compared with the case of deflection of light passing through a medium with a variable index of refraction. In a space in which the time coordinate is a Gaussian coordinate (see Chapter 4), the A_{4i} components $(i = 1, 2, 3)$, of the metric tensor vanish, and the line interval may be split into two parts: one including time differential only, $A_{44}c^2dt^2$, and the other including spatial differentials only. Denoting the second part dl^2, we write:

$$dl^2 = \sum_{i=1}^{3} \sum_{j=1}^{3} A_{ij} dx_i dx_j$$

and the interval can be written:

$$dS^2 = A_{44}c^2dt^2 - dl^2.$$

Here dl^2 is the squared coordinate length, which is the squared spatial interval between two events. If the interval is a world line of some object, we can divide it by dt^2, and, after reorganizing terms, we find:

$$\left(\frac{dl}{dt}\right)^2 = A_{44}c^2 - \left(\frac{dS}{dt}\right)^2.$$

(8.21)

The last expression is the squared coordinate velocity of the object whose world line equals to the line interval. (A coordinate velocity is the one measured in that particular coordinate system.) If we know how to calculate dS^2, we can determine the velocity of the object as measured in the certain coordinate system. For light, the interval always vanishes. Hence the coordinate velocity of light is given by:

$$\bar{c} = \frac{dl}{dt} = c\sqrt{A_{44}}.$$

(8.22)

where $\bar{c}$ is the light velocity as measured in the certain coordinate system. In a flat space $A_{44} = 1$, and light velocity is c. In a gravitational field, $A_{44} \neq 1$, and the velocity of the light measured is smaller than c (in the same way that the velocity of light in a medium whose index of refraction differs from unity is smaller than c).

We have an expression for the coordinate velocity of light in a curved space (a space influenced by gravity)[3]. The term $1/\sqrt{A_{44}}$ is the index of refraction of this space, and it can be used to calculate the trajectory of light according to Fermat's principle, as explained in Chapter 7. The results of using this method of calculation will be identical to those obtained in equation 8.19.

By using the refraction index of the medium, we arrive at the concept of gravitational lenses, which focus electromagnetic radiation (light, radio waves, etc.) coming from distant objects when passing through a high intensity gravity region. Such phenomena were observed when it was found that two different radiation sources located at a huge distance from us have identical spectra. The spectral properties of a source (its spectral distribution) characterizes the source uniquely. The spectral distribution depends both on the chemical composition of the source and on the thermodynamic state in the source. Added to these are the contributions of the Doppler shift due to the spatial motion of the source. Hence, each source has its characteristic spectral distribution, which might serve as the "fingerprints" of that source. The chance that two different sources might have identical spectral distribution is very remote. The explanation of the phenomenon observed—that two sources seemed to have the same spectral distribution—is that the two sources observed are actually two images of one and the same source. The two images are formed because a strong gravitational region resides between us and the distant source, and this gravitational field causes deflection of the radiation on its way towards us. The influence of the gravitational region causes focusing of the radiation; hence, two light rays which left the source and passed on two sides of the gravitational field were bent inward, and reached our telescope from different directions. Figure 8.4 shows the suggested trajectories of the two rays.

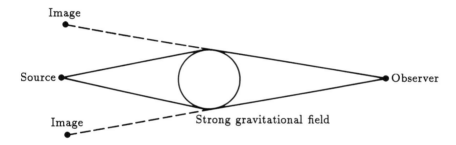

FIGURE 8.4.   Gravitational lens.   Light rays which left the source are bent on passing through a strong gravitational field. As a result, the rays focused and reach the observer as if they were emitted from different objects.

The situation presented in Figure 8.4 is an ideal one; the earth (the observer), the deflector (the strong gravitational field), and the source are exactly on the same line, and the images are symmetric with respect to this line. In most actual cases, the situation is not so ideal, and the images are not symmetric. In 1936, Einstein[4] had calculated what will be observed in an ideal situation such as the one shown in Figure 8.4, and he showed that, due to the symmetry along the axis earth–deflector–source, the images will be observed on a circle around the deflector and form a ring. At the end of his article, Einstein commented that the probability of finding such a situation in nature is practically zero, and the calculations have theoretical significance only. When our observations expanded over wider ranges of the electromagnetic spectrum, and also included radio wavelengths, several cases of gravitational lensing were observed, and people searched for the ideal case of "Einstein's ring," as it was called. The cover picture of this book is a wonderful picture of Einstein's ring. (It was taken by Jacqueline Hewitt and B. Burke from MIT, and Ed Turner from the Princeton University.)[5]

c. Red (or blue) shift of light due to gravity

This effect was already mentioned and explained in Chapter 6. The formula for the calculation of this shift is:

$$\frac{\lambda_2}{\lambda_1} = \sqrt{\frac{(A_{44})_2}{(A_{44})_1}}. \tag{8.23}$$

This formula is correct when the emitting atom is at rest. Yet in nature, atoms rarely stay at rest. The problem in measuring this effect in natural sources, such as radiating atoms in the sun or in white dwarf stars, is that the thermal random velocities of the atoms are high and they contribute to the Doppler shift of the radiation. Because these velocities are random, their contribution to the red or blue shifting is random too, and the spectral lines obtained in the observations are diffuse and broad, causing difficulties in the calculations. Moreover, the light emitted has momentum, and according to the law of momentum conservation, the emitting atom should gain momentum equal in magnitude and opposite in direction to the momentum of the light. The emitting atom recoils and this effect adds another shift to the wavelength, thus reducing further the accuracy of the measurements. Observations of gravitational red shift in astronomy are quantitatively unsatisfactory, although qualitatively they show the effect clearly.

More accurate results may be achieved in a controlled experiment in a laboratory, although the precision required in much higher. In an experiment performed in 1960 by Pound and Rebka (mentioned in Chapter 6), the potential difference was generated by a height difference of 22.5 meters between the emitting atom and the absorbing atom. Such an experiment requires an accuracy of 10^{-15}. ($\Delta\phi/c^2 = g \cdot l/c^2 = 2.5 \times 10^{-15}$). This degree of accuracy was achieved by the use of atoms of Fe^{57}, both as emitting and absorbing atoms. The participating atoms were imbedded in large crystals, which served as absorbers for the extra momentum of the recoil, as well as for preventing the emitting and the absorbing atoms from the random thermal velocities. Thus the influence of the two main disturbances was reduced drastically. In the experiment, the light was emitted upwards, so they expected a red shift of the light to occur. The crystal including the emitting atom was located on a large wheel whose velocity could be adjusted. It was arranged so that the radiation was emitted when the motion of the emitting atom was upwards. This motion in the direction of the radiation should cause a blue shift to the spectrum, which could compensate for the gravitational red shift. By adjusting the velocity of the rotating wheel, they reached a state in which the blue shift due to the velocity exactly compensated for the gravitational red shift. The wavelength of the absorbed radiation was equal to the emitted wavelength. The velocity of the source on the rotating wheel can be calculated precisely, and this yielded the Doppler shift expected, and from this the gravitational red shift could be calculated. The accuracy of this experiment was very high, and the error expected was a few percent of the predicted effect. The result fitted the theory very nicely.

It is worth noting that the shift of the wavelength of light due to gravity is not a test of the validity of Einstein's equations. It is rather a test of validity of the principle of equivalence, and for the value of the component A_{44} in the metric tensor. As explained in Chapter 6, the principle of equivalence is at the basis of the GTR, but other theories of gravitation which include the principle of equivalence can account for this effect, even if they do not include the complete theory of the GTR.

Another prediction which became available for measurements recently is the measuring of the reduced velocity of radiation when passing through a gravitational field. This velocity reduction is due to the "refraction index" of the curved space created around a massive object. The effect is measured by measuring the time needed for light motion when the light trajectory is very close to the sun. The signal was sent from the earth to "mirrors," and the time was measured for the return trip. Venus and Mercury were used as mirrors for part of the observations, and active reflectors located on spacecrafts were used as mirrors in other parts of the observations. The reduction of the velocity is of the order of magnitude of GM/c^2r, which is a few microseconds, while the time needed for the radiation for a return trip is a few hundred seconds. The predicted velocity reduction was observed clearly, with the accuracy of a few percent.

8.4.   Schwarzschild Radius

A close inspection of the components of the metric tensor of Schwarz-schild's exterior solution shows that there is a value of r which may cause difficulties. The value $r = 2GM/c^2$ yields $H \to 0$; $B \to \infty$. This radius is called Schwarzschild's radius. The behaviour of B and H shows us that the coordinate system we used does not suit the description of the physical space at this point. The region interior to Schwarzschild's radius is called a "black hole," for reasons we shall go in later. We can study what happens to an object approaching this region by following an object proceeding in a free-fall motion towards the center of the mass. The information we receive about an object reaches us by the radiation emitted by the object. It is clear that the spectrum of the radiation emitted by the object is red shifted as the object comes closer to the gravitational field. There will also be a red shift due to the Doppler effect, but this can be calculated and subtracted, and we shall be left with the gravitational red shift alone. According to our calculations in Chapter 6, we have:

$$\lambda_o = \frac{\lambda}{\sqrt{1 - \frac{2GM}{c^2 r}}} \tag{8.24}$$

where λ is the wavelength emitted by the object and λ_o is the wavelength absorbed by a distant observer (located where $\phi = 0$). We find that when the object proceeds towards the center of gravity and emits radiation with a constant wavelength λ, the wavelength of the absorbed radiation will become longer and longer. When the object reaches Schwarzschild's radius, the wavelength will tend to infinity, which means that the radiation frequency tends to zero. The radiation we receive from the falling object will be shifted more and more towards the red end of scale, until the object totally disappears when it reaches Schwarzschild's radius.

Another factor will influence the radiation emitted. The coordinate velocity of the radiation becomes lower as the gravitational field becomes stronger because the refraction index increases: $\bar{c} = c\sqrt{1 - (2GM/c^2 r)}$. Hence, the velocity of the radiation emitted from the falling object (which should inform us about what happens there) is reduced, and the time it needs to reach us becomes longer and longer. The radiation emitted when the object is located on the Schwarzschild's radius will not reach us at all. It is found that the time needed for the object to reach Schwarzschild's radius will seem to be infinite to an exterior observer.

However, the observer located on the object free-falling into the black hole will measure the time according to his clock. This observer is located in an inertial system. He will measure the time needed to reach Schwarzschild's radius and will find finite time. But the exterior observer will have no way to receive this information from the interior observer. Two factors combine to hide the trespassing of Schwarzschild's radius from the exterior observer: the red shifting of the wavelength which end with an infinite wavelength for the emitted radiation, and the slowing of the motion of the radiation which ends with total stop of the emitted radiation at this point.

It is interesting to study the feeling of the observer located on the object free-falling into a black hole. The object is free-falling, hence its system is an inertial system for the observer who moves with it, and no forces are felt in this system. But the situation is not so convenient as it might seem at first glance. As the object approaches Schwarzschild's radius, the gravitational field becomes stronger, and its rate of change with location becomes stronger. The differences in the field strength between nearby points whose distance from the gravity center is different becomes very large. These differences cause different curvature of space at different points, and these differences become very strong. These differences in the curvature create forces called "tidal forces" (these are the forces which cause the ocean tides on earth). A point object proceeding in a free-fall will not feel the tidal forces. But any real object has a certain volume, and the differences in the field strength between different points in the object will create strong tidal forces which will probably tear apart any real object whose volume is not zero. Around the points where $r = 0$, tidal forces are infinite.

8.5. Gravitational Collapse

An object whose entire mass is already interior to Schwarzschild's radius is said to be gravitationally collapsed. Schwarzschild's radius is also called the gravitational radius. At this point $A_{11} = -B$ approaches infinity; the coordinate system we used 'crashed' and cannot be used any more. In order to be able to calculate what happens to objects inside the black hole, more sophisticated methods should be used, such as transforming to another coordinate system which will not crash on passing through the gravitational radius. However, the consequences of such calculations are that no material object, not even a photon, can pass through the Schwarzschild's radius in the outward direction. Only inward travel is possible. Therefore this region is called a black hole. It absorbs everything—matter and radiation—and nothing can escape.

We may try to understand this behaviour by studying the properties of H and B, the components of the metric tensor:

$$A_{44} = H = A_{tt} = 1 - \frac{2GM}{c^2 r}$$

$$A_{11} = -B = A_{rr} = \frac{-1}{1 - \dfrac{2GM}{c^2 r}}$$

Outside the gravitational radius, t is a timelike coordinate ($A_{44} > 0$) and r is a spacelike coordinate ($A_{11} < 0$), as we expect them to be. Inside the gravitational radius, $2GM/c^2 r > 1$, the signs of A_{44}, and A_{11} are inverted, r becomes a timelike coordinate, and t becomes a spacelike coordinate; What does this signs inversion cause?

When we included time as a fourth coordinate in the space-time system, we treated it mathematically in the same way we treat the spatial coordinates. But we still did not change the intrinsic difference between the two kinds of coordinates. Along a spatial coordinate axis, one can move in the negative direction as well as in the positive one, according to the physical situation. But one cannot move in the two directions along the time axis. Along this axis one can move forward only. No object can move backward in time. Actually this is the physical content of the definition of a timelike interval. All the causality system (relation between cause and consequence) is based on this property of time, and a backward motion in time would destroy the whole causality system.

When r becomes a timelike coordinate, one can move along its axis in one direction only, and this direction is towards $r = 0$. Once an object has passed through the Schwarzschild's radius, its fate is determined—it will end at $r = 0$. A spacecraft which happened to enter this region may use all its engines' power in an attempt to escape this fate, but at most it can slow down its approach to $r = 0$. The timelike character of the r coordinate in this region will not allow it to retrace its path. The influence on the t coordinate will be the opposite. Outside the gravitational radius, t is the timelike coordinate, where motion is possible in one direction only, to the future. But for an observer located inside the gravitational radius, t has converted to a spacelike coordinate, and the time arrow can point at any direction. An observer located inside the gravitational radius may watch a clock located outside the gravitational radius. The direction of motion of that clock (as observed by the interior observer) depends upon the state of motion of the observer. There might be situations in which the outside clock will seem to move backward. However, the clock located with the inside observer will always move forward.

What is the Schwarzschild's radius for the objects we know? If we insert into the term $r = 2GM/c^2$ the numerical values for the sun, we find that Schwarzschild's radius of the sun is about 3 km. This means that if we press the total mass of the sun into a ball with a radius of 3 km, then it will behave like a black hole. (The present radius of the sun is about 700,000 km.) Schwarzschild's radius for the earth is about 0.85 cm. In both cases, Schwarzschild's radius is located very deep inside these objects. As the exterior Schwarzschild's solution is valid for the region void of matter, the objects mentioned here are very far from gravitational collapse. The most condensed objects known to date are the neutron stars. Their radius is several times larger than their Schwarzschild's radius. The fact is that none of the objects we know is gravitationally collapsed. Evidently, an object whose actual radius is smaller than its Schwarzschild's radius cannot be observed by direct observations. The chances to observe the existence of a black hole are only in finding phenomena which indirectly show the existence of such an object.

8.6. Effective Potential

A better insight into the situation of a black hole may be achieved by calculating the energy of a test particle in the field of a large concentration of mass, M. We shall calculate it both by the Newtonian calculation and the relativistic one. We shall work with specific quantities (per unit mass) for the energy, the momentum and the angular momentum. By the

Newtonian calculation we have:

$$E = K - \frac{GM}{r} + \frac{L^2}{2r^2}.$$ (8.25)

K is a specific kinetic energy connected with radial motion, and all the other symbols have the same meaning as in equation 8.10. The term $L^2/2r^2$ includes the kinetic energy connected with the rotational motion. (For example, when an object moves in an exact circle, $L = rv$; $L^2/2r^2 = v^2/2$, which is the specific kinetic energy of the rotational motion and $K = 0$.)

On multiplying equation 8.10 by $L^2/2$ and comparing it with equation 8.25, we find that

$$K = \frac{L^2}{2} \cdot \left[\frac{d}{d\varphi} \left(\frac{1}{r} \right) \right]^2.$$

In a central field, as we have here, the angular momentum is a conserved quantity, and it is a constant of motion. The value of the angular momentum is $L = \omega r^2$. Hence $L^2/2r^2 = \omega^2 r^2/2$, which is equal to the centrifugal potential (equation 7.9).

Let us denote the last two terms in equation 8.25 "effective potential," V. The effective potential includes the gravitational potential $-GM/r$, which is negative, and the centrifugal potential, $L^2/2r^2$, which is positive (a repulsive potential), and it fall with distance like $1/r^2$. Due to the different dependence of the two terms on r, the behaviour of the effective potential changes with distance. At large distances, the term which depends on $1/r^2$ is negligible, and the gravitational potential is dominant. Hence, at large distances, the test particle feels mainly the gravitational attraction towards the central mass. When the test particle comes closer, r becomes smaller and the relative weight of the centrifugal potential increases until it dominates the effective potential.

In this case the particle is repelled by the centrifugal potential from the central mass. The sign of the total energy, E, determines whether the particle will be repelled to infinity or if it will be bounded in a closed orbit around the central mass. In Figure 8.5, the effective potential, V, is drawn versus the distance from the center of the mass. Two possible orbits are drawn: l_1, l_2. For l_1 the total energy is negative, and the particle will stay

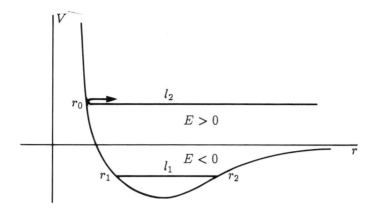

FIGURE 8.5. Effective potential calculated by Newtonian calculation. The orbit of an object whose total energy is negative is l_1, and it moves in a closed orbit between radii r_1, r_2. The orbit of an object whose total energy is positive is l_2, which, after reaching a minimal approach r_o, will rebound to infinity.

in a bound closed orbit where the maximal and minimal distances in this orbit are r_2, r_1 (elliptical orbit). For l_2 the total energy is positive. This means that the energy which represents repulsion, $K + L^2/sr^2$, is bigger than the absolute value of the gravitational attractive energy. The particle will reach a minimal approach, r_o, and then rebound to infinity. Except for the case of a head-on collision, in which $L = 0$, the graph of the effective potential goes to infinity when $r \to 0$. Hence, for each particle which does not move on a head-on orbit, there exists a certain distance, r_o, at which the centrifugal force overcomes the gravitational force, and the particle approach is limited to this distance (which may be calculated when the initial conditions are given). At an infinite distance from the central mass, the effective potential approaches zero, and the total energy is equal then to the kinetic energy, $E = K$.

In the relativistic calculations, the equation of the orbit is obtained by the calculation of the geodesic orbit of a free particle in the curved space around the central mass. The equation obtained is:

$$E^2 = K^2 + \left(\frac{-2GM}{r} + \frac{L^2}{r^2} - \frac{2GML^2}{c^2 r^3} + c^2 \right) \cdot c^2. \qquad (8.26)$$

The right wing of this equation includes K^2 and a second term that we shall denote by V^2, where V is the effective potential. (In some books the authors denote this term by V. For additional details, see "Gravitation" by Misner, Thorne and Wheeler[2].) We may compare this equation with equation 8.11. The energies here appear in the second power. This is characteristic for relativistic equations which connect energies with dynamic variables.

On comparing equation 8.26 with equation 8.25, we see in the effective potential two additional terms, the two last ones. One of them is a constant c^2, and the second is L^2 multiplied by the term $-2GM/c^2r^3$, which we already know from the relativistic equations we have dealt with. The constant c^2 reflects the fact that in relativistic calculations we include the rest energy of the particle in the energies sum ($E = mc^2$). As we deal with specific energies here, the rest energy per unit mass is c^2. The addition of this constant to the energy equation does not change much the character of the variables, as energies are usually defined up to an additive constant. The terms which determine the character of the equation are energy differences, not absolute values of energies. The influence of this term is to move the whole graph of the potential upward so that when $r \to \infty$, the effective potential approaches c^2 (instead of 0).

The second term, which include $-2GML^2/c^2r^3$, represents the difference between the Newtonian equation and the relativistic one. As this term falls with distance as $1/r^3$, its contribution to the effective potential at large distances is negligible. The appearance of c^2 in the denominator makes this term very small numerically. Only when the particle reaches very small distances from the center ($r \to 0$) it does dominate the effective potential.

In Figure 8.6 the behaviour of this potential is shown versus r. At large and medium distances, the behaviour of the potential is similar to that of the potential in the Newtonian equation. At large distances the dominant is the gravitational potential, and the particle is attracted towards the central mass. When the particle approaches the central mass, the relative weight of the centrifugal potential increases and the effective potential is a repulsive potential. But on a further approach, the third term dominates the potential. This term is negative, it represents attraction, and the particle is again attracted towards the central mass. The maximum in the graph of the effective potential shows the distance at which this transition occurs.

Here too we have an orbit l_1, whose total energy is lower than c^2. This

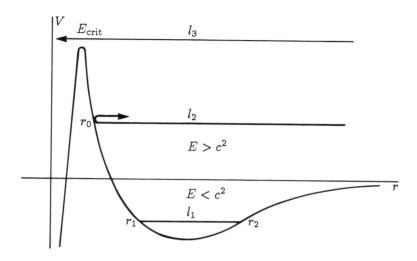

FIGURE 8.6. Effective potential calculated by relativistic calculation. The orbits l_1, l_2 correspond to those appearing in Figure 8.5. In orbit l_3, the total energy is higher than the critical energy in which the effective potential becomes an attractive potential.

orbit is elliptic, between the distances r_1, r_2. The orbit l_2, has total energy higher than c^2, and this is the orbit of the particle which reaches a minimal approach and rebounds to infinity. However, here we have a third kind of orbit. The orbit l_3 has energy much higher than c^2. The particle passes over the critical point, at which the third term is dominant. When the particle reaches this point, the repulsion turns into attraction, and it will continue its motion towards the center without return.

The additional term, $-2GML^2/c^2r^3$, contributes along the whole range of distances. But at large distances its contribution is minute and can usually be neglected, except for minute deflections from the orbits calculated according to the Newtonian equation, such as the precession of the perihelion of the planets. For very low values of r, its influence becomes critical, and it causes the absorption of the particle by the central mass. The limit for the critical distance depends on the energy. In the Newtonian calculation, we have found that the higher the energy, the stronger it will rebound. In the relativistic calculation, we find that very high energies cause a situation where there is no rebound at all. This behaviour reflects the gravitational properties of the energy, which cause the non-linear character of the GTR. In the Newtonian equation, the magnitude of the energy did not change the gravitational attraction because only the rest mass of the particle participated in the gravitational interaction. In the relativistic equation, the kinetic energy connected with the angular momentum L contributes to the gravitational attraction through the term $-2GML^2/c^2r^3$. The contribution of the angular momentum to the gravitational attraction may be seen clearer if we write together the second and the third terms in the effective potential:

$$\frac{L^2}{r^2} - \frac{2GML^2}{c^2r^3} = \frac{L^2}{r^2}\left(1 - \frac{2GM}{c^2r}\right)$$

This way of writing clearly shows that the centrifugal potential L^2/r^2, whose contribution usually causes repulsion, turns to be an attractive potential in the region of Schwarzschild's radius. Since there also always exists the regular gravitational potential, the critical point at which the negative part of the effective potential overcomes the kinetic energy is outside the Schwarzschild's radius.

8.7.   Black Hole

By its definition, a black hole is an object whose radius is smaller than its Schwarzschild's radius. Matter and energy which reach the gravitational radius are absorbed, and their mass is added to the mass of the black hole. As a result, its gravitational radius increases too. Since no mass can leave the black hole, its fate is to grow and never become smaller.

Recently it was found by Bekenstein and Hawking that black holes can lose energy by certain processes. The rate of these processes is inversely proportional to the mass of the black hole. Small black holes (whose masses are about ten orders of magnituide smaller than the earth's mass) emit energy at a significant rate. The mass of the black hole reduces, and the process is accelerated until the black hole is totally "evaporated." In large black holes (one solar mass and above), the rate of energy loss is negligible, and they will probably continue to grow by absorption of more masses. There is no justice in nature: the small will get smaller, and the big will get bigger.

We dealt with the black hole as an isolated system and with a point particle attracted by the black hole. In a continuous distribution of mass, such as plasma in a star or a multi-particle system like a dense star cluster, there exists the possibility of the creation of a black hole at the center of the system. The gravitational force (which is the relevant force over large distances in the universe) drives the system towards denser configurations by the mutual interaction between the constituents of the system, or alternatively, by the space curvature induced by the presence of matter. Were gravitation the only force in nature, all the systems would develop toward total collapse. In a multi-particle systems where frequent encounters between particles exist, the angular momentum and energy of each particle separately are not conserved. The laws of conservation are valid for the system as a whole, and we deal with the properties of the system as a whole.

The kinetic energy and the angular momentum of the particles act against the collapse. These entities, and all the other interactions in the system which counteract the collapse, are represented by the pressure in the system. The fate of the system is determined by the balance between gravity (which tends to contract the system) and the pressure (which counteracts the contraction). In a spherical symmetric system like a star, we see gravity as pushing towards the center, while the pressure acts outwards, opposite to gravity. The force created by the pressure is due to the gradient of the pressure from the center outward. At the center the pressure and the density gain their maximal values. When the star contracts, the central density increases. Once the density at the center increases to such a value that at some point r the relation $2GM(r)/c^2r > 1$, ($M(r)$ is the mass included in a sphere of radius r), this region turns into a black hole and behaves like a black hole with respect to the material around it. Increasing density works towards increasing gravity and earlier formation of a black hole.

On the other hand, increasing density causes an increase in the pressure which counteracts the collapse. The winner in this race is determined by the detailed form of the dependence of the pressure on the density. The steeper this dependence, the stronger are the forces which counteract the collapse. If the dependence of the pressure on the density is weak, it will not increase fast enough to stop the collapse. The equation which describes the dependence of the pressure on the density is called "the equation of state." The character of the equation of state is very important for the study of the development of dense systems. The equation of state reflects the micro-interactions between neighbouring particles while gravity reflects the macro-physics of the entire system. The balance between gravity and the pressure gradient is the link which connects macro-physics and micro-physics, or global considerations with atomic and molecular physics. On discussing evolution of astronomical objects, the details of the equation of state in different situations should be considered. We shall return to part of them in the next chapter, when we deal with relativistic phenomena in astronomy.

8.8. Interior Schwarzschild's Solution

The interior solution of Schwarzschild deals with properties of space inside the mass distribution. The major difference between the exterior and the interior solution is that inside matter the energy-momentum tensor, **T**, does not vanish. In this region too we assume spherical symmetry and the existence of a static solution. Hence, the energy-momentum is quite simple (equation 7.14). However, Einstein's equations can be solved analytically only with another assumption, namely, that the density is constant over the entire mass and that it is not connected with the pressure. This assumption is very unreal for the matter we know in nature, and hence the solution obtained with this assumption yields only a very general picture of the real configuration in matter. We shall not go into detail about this solution. On the other hand, the exterior solution of Schwarzschild yields quite a good description for the space around an isolated object.

Comments and References

1. Actually, each planet forms with the sun a pair of rotating objects which rotate in elliptical orbits around their common center of mass with the same frequency. The relative distance of each object from the center of mass is inversely proportional to its mass. Because of the big ratio of the solar mass to the mass of the planet, the center of mass is deep inside the sun, close to its center. The motion is approximately a motion of the planet in an elliptical orbit around the sun. All the considerations about the orbit are correct for the "orbit" of the entity denoted the "reduced mass."

2. Misner, C. W., Thorne, K. S., Wheeler, J. A. *Gravitation.* Freeman and Co., San Francisco, 1973.

3. Moller, C. *Theory of Relativity.* Oxford University Press, London, 1966.

4. A. Einstein, *Science,* 1936 **84,** 506.

5. J. N. Hewitt, E. L. Turner, D. P. Schneider, B. F. Burke, G. I. Langton, C. R. Lawrence, *Nature,* 1988, 333, 537.

Cosmological Solutions

In large scale systems the only significant interaction is gravity. One of the reasons for this is that nuclear forces are very short range. They tend to zero very fast with distance, so that on scales of distances larger than those of atomic nucleus, they practically vanish. The electromagnetic interaction, which is long range, depends on the excess charge of the interacting objects. We have two kinds of charges (positive and negative) and two kinds of interactions (attraction and repulsion). The interactions in any system depend on the excess of charges on the different constituents of the system. If no charge excess exists, the equal number of charges of the two kinds balance each other and the system appears as a neutral object and is not active electrically. On the average, the matter in the universe is electrically neutral. The repulsive and attractive interactions balance each other and no electromagnetic interaction exists between large accumulations of matter.

We did not mention the electromagnetic radiation, which is actually also an electromagnetic interaction. The nature of this radiation is of periodical oscillations in the electric fields created by changes which take place at a charged object which serves as the source. These oscillations in the electric fields proceed in space at the velocity c, and their magnitude falls with distance as $1/r$. The power transferred by these oscillations is inversely proportional to the squared distance from the source. In the present state of the universe, the energy density of the radiation is very low relative to the energy density of matter. Hence the influence of radiation may usually be ignored. In states in which the energy density of the radiation is high, it can influence the dynamics of objects. For example radiation pressure can impart acceleration to material objects. States in which the radiation pressure should be consider ed exist close to stars whose luminosities are very high, and they did exist at the first epoch of the universe when the universe volume was much smaller and the energy density was very high.

In gravity we have only one type of charge (the mass) and only one kind of interaction (attraction). Hence, although gravity is the weakest interaction in nature, when we deal with huge amounts of matter, its influence accumulates. In such situations gravity is the only interaction that remains over large distances. It is natural to expect that a theory of gravity, as the GTR is, will be able to describe a general picture of the universe.

9.1.   The Cosmological Constant

The first attempt to solve Einstein's equations for the entire universe was done by Einstein himself. He tried to find a static cosmological solution. The assumption that there should exist a static cosmological solution caused difficulties.

As was described in Chapter 8, a material system which is subject to gravity will tend to contract, due to the interaction between its constituents. From observations about the universe around us, we find enormous distances between the stars, the pressure in the universe is actually nil, and there is nothing that can prevent the systems of stars and galaxies from collapsing. In such a situation, a static cosmological solution is impossible. In order to obtain a static solution, Einstein added a constant to the equations denoted by Λ, and known as the "cosmological constant." The role of this constant is to prevent the model from contracting, thus enabling the static state to exist. The cosmological constant represents a force which acts in the opposite direction as gravity, and it does not depend upon matter, but on space. Equation 7.13 becomes:

$$G_{ij} + A_{ij}\Lambda = JT_{ij}. \tag{9.1}$$

Solving the equations reveals that the magnitude of this constant is the inverse of the squared "radius of the universe." This radius is estimated to be about twenty billion light years. Hence the value of Λ is very small, and its influence is significant only over inter-galactic distances. When considering the solar system and even the entire Galaxy, the influence of this constant cannot be observed.

But it turned out that such a solution is not stable against minute perturbations in the density. Suppose that we have such a state in which the force represented by the cosmological constant just balances gravity. This is the static solution. Now, if a small increase in the density occurs (which means shorter distances between the constituents of the system), then the gravitational attraction will increase too, and it will overcome the balancing constant. A contraction will follow which will increase the density. This will enhance gravity further and so on, and the contraction will continue until there is a total collapse. The cosmological constant, whose influence does not depend on matter, cannot prevent a contraction once it started. On the other hand, if a small decrease in the density occurs in a state of equilibrium, gravity will become weaker and the force represented by the cosmological constant will cause an expansion. This expansion will cause further decrease of the gravitational force, and the expansion will eventually proceed indefinitely.

We expect that natural systems which reached their present state through physical evolution will be stable against small changes in the physical parameters. The evolutionary path in nature is not predetermined. It is actually chosen at each moment between a wide range of possibilities. The choice is made according to the balance of the factors at each instant. This balance is a dynamic one and not static, and at each momentary change it is created anew. A stable system is one in which small changes in the physical parameters will give rise to forces acting in a direction opposite to the changes, which will drive the system back to the original state. Such a mechanism is called "negative feedback" and its existence stabilizes the system. Only systems which include some kind of negative feedback mechanism can survive for an appreciable amount of time. In a static cosmological model based on the existence of the cosmological constant, we actually have a positive feedback mechanism which destabilizes the system. It is very unlikely that our universe (which we believe to have existed for billions of years) could survive in such delicate and unstable balance between gravity and the cosmological constant.

Thus, the static solution, although consistent mathematically, was not satisfactory physically and time-dependent solutions were investigated. Later it was found and established by observations that the entire universe is in a state of expansion, and these observations tempted the search for time-dependent solutions which would agree with the expansion observed.

When the idea of a static cosmological solution was abandoned, there was no need for the cosmological constant in the equations, and Einstein himself advised against its use. The simple cosmological models suggested today do not include the cosmological constant. On the other hand, alternate use was found for it. In a search for alternate theories for the GTR, and in efforts to treat special problems in relativity, the cosmological constant is still in use.

9.2.   Friedman's Model

The cosmological solution to Einstein's equations is an interior solution, as we are inside the matter distribution. We shall describe here the solution proposed by Friedman in 1922, which is an expanding model of the universe. We assume homogeneity and isotropy of the universe, hence it is again convenient to work in a spherical polar coordinate system. The matter fills space homogeneously and all points in space are equivalent. The density of matter is assumed to be the same over the entire universe, which means that we deal with an average density, although we know that there are local concentrations of matter in the form of galaxies and stars. The matter around us is far from being homogenous. On the scale of the solar system, we know that the sun holds most of the matter in the system with tiny material points around it in the form of planets moving in an almost empty space. The Galaxy is an ensemble of about 10^{11} stars in a disc with a diameter of 100,000 light years, and the distance to the nearest similar galaxy is about two million light years. But when we look to larger distances and collect information from distances of billions of light years, we find that the matter in the universe is distributed quite homogeneously, and, on this scale, the concentrations are considered as local deviations from the average density and do not change the overall picture.

The assumption of homogeneity means that matter density is uniform at each moment over the entire universe, but it may change in time. The general form of the interval in such a model is:

$$dS^2 = Hc^2dt^2 - Bdr^2 - Dr^2d\Omega^2 \tag{9.2}$$

where D, B, H are the diagonal components of the metric tensor, and they may be functions of time and location. Due to the isotropy, there is no dependence on the angular coordinates, and the spatial dependence is on r only. As we deal with an interior solution, the energy-momentum tensor does not vanish and we have to consider the form of this tensor.

9.3. Co-Moving System

In the present case we can use a coordinate system called a "co-moving" system. Such a system is attached to the bulk of matter and moves with it. In such a system the average motions of the matter particles vanish. The bulk motion of matter is characterized by the motion of the coordinate system, and only internal motions such as motions of particles relative to the bulk of matter can be observed in such a system. For instance, if we attach a coordinate system to a moving train, then the motion of the train coincides with the motion of the system, and inside this system we can observe only motions of objects which move relative to the train. The coordinate system is a co-moving one and the train is at rest in this system. When we treat a cosmological model, there is no significance in talking about a motion of the whole system—neither a translational nor rotational one—since we can relate this motion to no other material system.

We can, however, talk about expansion or contraction of the model either of which might change the internal distances between points in the universe. In such a case, the co-moving system is expanding or contracting with the universe, and each material point in the universe (a galaxy, for instance) will have a fixed location coordinate although its distances from all other galaxies will change. The coordinates in this case serve as markers, denoting the positions of objects in the coordinate system. The coordinate difference between two objects at rest in the system will remain unchanged although the distances between objects in the system might change due to the expansion or contraction of the system. In order that the coordinate differences will play a role as distance markers, we have to multiply them by the corresponding components of the metric tensor.

For instance, let us take a distance between two points, P_1, P_2, whose angular coordinates are the same: $(\theta_1 = \theta_2; \varphi_1 = \varphi_2)$. The coordinate difference between these two points includes radial difference only: $(d\theta = d\varphi = 0; dr = r_2 - r_1)$. The differential dr shows only coordinate difference between the two points. In order to obtain from the differential dr the distance dl, we have to multiply it by $\sqrt{B} = \sqrt{-A_{11}}$ to obtain: $dl = \sqrt{B}dr$. Here, dl has a meaning of a measurable distance. Thus, in a co-moving system, matter is on average at rest, and time evolution is observed through the time evolution of B.

This situation may be demonstrated for a two-dimensional space by a rubber balloon which can be inflated or deflated at will. Suppose we take the balloon when it has a certain radius and mark a net of longitudes and latitudes on its surface. Each point on the balloon has a definite coordinate location given by its longitude and latitude. Let us inflate the balloon to a radius larger than its initial one. Each point on the balloon remained in its initial location relative to the coordinate net, although the whole balloon was inflated and increased in its dimensions. The coordinates of each point remained the same, although its location in space definitely has changed. The distances between each two points on the balloon also changed, although coordinate differences between the points did not change. Such a coordinate system is a co-moving system. If we want to study the temporal behaviour of points on the balloon's surface, we have to study the time evolution of the balloon's radius (which is a part of the metric tensor components.) If we want to measure distances on the balloon's surface, we multiply coordinate differences by the corresponding component of the metric tensor, which is R for $d\theta$, and $R\sin\theta$ for $d\varphi$. (R is the balloon radius.) If we want to mark points inside the balloon, we have to add a third coordinate, r. This coordinate marks "coordinate distance" from the center, because they mark a distance relative to the radius of the balloon. If R is measured by standard length units, then the coordinate r (which marks proportional parts of the balloon's radius) receives a meaning of length when it is multiplied by R. ($l = r \cdot R$). Here R is a time-dependent quantity; it increases or decreases with the expansion or contraction of the balloon. Hence we write $R(t)$. Here r is a co-moving coordinate. The coordinate difference dr should be multiplied by $R(t)$ which transforms the coordinate differences into measured distances.

In such a case, all the time evolution of the system as a whole, as well as its curvature, is included in the components of the metric tensor. This example can be pursued by analogy to the description of the universe, but it must be noted that this description concerns time evolution of the bulk of matter. Local movements of specific mass points should be studied locally.

In the co-moving system, the matter is at rest. Hence, from the four-vector of the momentum we are left with the fourth component only, the component which includes the mass density. Due to the isotropy assumed to exist in the universe, only normal forces can exist, and the stress tensor includes only the diagonal components which represent the pressure. We can treat the matter as an ideal fluid and use the energy-momentum tensor given in equation 7.17 or 7.18.

9.4.   The Metric Tensor of Friedman's Model

Let us return to the components of the metric tensor, D, H, B , given in equation 9.2. From the field equations[1] we can derive an equation which presents the dependence of H on the spatial coordinates. The equation is:

$$\frac{H'}{H} = \frac{-2P'}{(\rho c^2 + P)} \tag{9.3}$$

where $P' = dP/dr$; $H' = dH/dr$. Here ρ, P, are the density and the pressure. Due to the homogeneity, the pressure is the same at all points in space, and hence $P' = 0$. The density is always positive, the pressure can be zero or positive (but never negative); hence, the denominator of the right wing of equation 9.3 is always positive. If $P' = 0$, then also $H' = 0$. This means that H does not depend on the spatial coordinates but on time alone. (In the exterior Schwarzschild's solution we had: $\rho = P = 0$; hence, the relation 9.3 could not be used there.) As mentioned in the preceding Chapter, a coordinate whose coefficient in the interval does not depend on other coordinate is called a Gaussian coordinate. In the present system, time is a Gaussian coordinate. As H depends on time alone, we can define a new time coordinate, t', which includes H $(dt' = \sqrt{H}dt)$, and thus obtain for the interval (dropping $'$):

$$dS^2 = c^2 dt^2 - B dr^2 - D r^2 d\Omega^2. \tag{9.4}$$

The assumption of the homogeneity of space implies that the whole spatial part of the interval will depend on time by the same form. Let us characterize the time dependence by $R(t)$, where $R(t)$ is a time dependent function which scales the universe. As for the spatial dependence of the spatial part of the interval, this dependence is on r alone. Thus we can define a new r which includes D, and we obtain for the interval:

$$dS^2 = c^2 dt^2 - R^2(t) \left(B(r)dr^2 + r^2 d\Omega^2 \right). \tag{9.4a}$$

The $B(r)$ appearing here is different from B which appear in equation 9.4.

We have to solve the equations for two variables: $R(t)$ which represents the dependence of the model on time, and $B(r)$ which represents the dependence of the model on the spatial coordinate. The general form of the equations is:

$$\mathbf{G} = J\mathbf{T} \tag{9.5}$$

where J is a constant $(J = -8\pi G/c^2)$, **T** is the energy-matter tensor, given in equation 7.17, and **G** is Einstein tensor, which includes the components of the metric tensor and their derivatives. The solution of the field equations for $B(r)$ is:

$$B(r) = \frac{1}{1 - Qr^2} \tag{9.6}$$

where Q is a constant of integration whose value and sign should be determined according to the boundary conditions of the model. It comes out that Q is the Gaussian curvature of the model. (In the numerator of equation 9.6 we obtained 1, as we used the boundary condition that $B(r) \rightarrow 1$ when the curvature tends to zero.) It is convenient to separate the definition of the sign and the value of Q by writing $Q = k/R_o^2$, where R_o is a constant whose dimensions are like those of r, and k is a pure number which can assume the values 1, 0, -1.

The sign of k tells us whether the curvature is negative, positive or vanishes. The value of R_o represents the radius of curvature of the model. The larger R_o is, the more flat the model is. Inserting $Q = k/R_o^2$, we obtain:

$$B(r) = \frac{1}{1 - \dfrac{kr^2}{R_o^2}}. \tag{9.6a}$$

When we substitute equation 9.6.a into the interval, we get:

$$dS^2 = c^2 dt^2 - R^2(t)\left(\frac{dr^2}{1 - \dfrac{kr^2}{R_o^2}} + r^2 d\Omega^2\right). \tag{9.7}$$

When the interval is given in this form, the properties of the model are expressed clearly. The spatial part of the interval includes a part which depends solely on space (the expression in the brackets), multiplied by a time dependent factor, $R(t)$, which serves to scale the model. The spatial part is constant, which means that the space curvature does not change during the evolution. The model as a whole can expand or contract according to the behaviour of $R(t)$.

In parallel to the example of the balloon, R_o represents the radius of the model, and the quantity r/R_o marks the relative position of a point r with respect to the entire system. We cannot relate to r (or to R_o) regular length units because the system is a co-moving one, which contracts and expands with the model. If we relate to R_o length units, they will be changing length units, which shorten or lengthen with the model. In parallel to our definition of coordinate time in equation 7.10 (dt), we define here a coordinate length dr. The real length (independent of the state of the model) measured in such a system is obtained when we multiply the coordinate length by the scale factor $R(t)$. The concept of length in this model will become clearer if we write equation 9.7 in the form:

$$dS^2 = c^2 dt^2 - R^2(t) R_o^2 \left(\frac{\frac{dr^2}{R_o^2}}{1 - \frac{kr^2}{R_o^2}} + \frac{r^2}{R_o^2} d\Omega^2 \right) \qquad (9.7a)$$

In this form each quantity in the brackets is dimensionless. The meaning of length is attributed to these quantities when they are multiplied by $R(t) \cdot R_o$, where $R(t)$ is a scale factor which changes with time, and R_o has the dimensions of coordinate length. Through this form it is clear that the product $R(t) \cdot R_o$ is proportional to physical distances measured in the model.

9.5.   The Constant k

How does the constant k influence the model? The constant k can assume three different values: 1, 0, -1. Let us study what properties of the model are obtained for each value of k. If $k = 0$, the spatial part of the interval in equation 9.7 will be: $(dr^2 + r^2 d\Omega^2)$. This form we already know, and it suits a spherical coordinate system in a flat space with zero curvature (as in the STR). This flat space is time dependent, and it contracts or expands with the scale factor $R(t)$. This is a very simple solution and it will be used as a limiting case when we study more complicated states.

The sign of k represents the curvature, whether it is positive or negative.

The second term in the denominator, kr^2/R_o^2, yields a measure to the magnitude of the curvature (the deviation from flatness). The form of this term shows that the measured curvature depends upon the location where it is measured, and it grows with the (coordinate) distance r. The properties of this model are based on the assumption that it is homogenous: all the points in the model are equivalent. Any point can be chosen to be the origin for the coordinate system, and, evidently, the most convenient choice will be to choose the observer's position as the origin from which all distances are measured. Each observer will find that for small r (which means in his close vicinity), the space is similar to a flat space. The further he looks up in space, the more curved the universe will seem to him. (This picture is based on the assumption of average density. Local concentrations of mass could change the picture around them significantly.)

In the case where $k = 1$, when r increases, the quantity $1 - r^2/R_o^2$ decreases, and B increases. When $r \to R_o$, the denominator of B tends to zero and $B \to \infty$. This puts an upper limit for r, which is $r = R_o$. This means that there exists a maximal possible distance between any two points in the universe. One can go away from a certain point only the maximal distance where $r = R_o$. Any attempt to move further will cause a decrease of the distance. Such a model is a closed model. A positive curvature is characteristic for a closed universe.

What is the physical meaning of this statement? Let us use again the example of the balloon. If we begin from a point P_1 on the balloon's surface, and move always with our back to P_1, our distance from P_1 will increase up to a certain point, which is the opposite to P_1 on the balloon's surface. If we continue to move from that point, no matter in what direction, our distance from P_1 will decrease. This is the meaning of a closed space in two-dimensional system, the balloon's surface. It is more complicated to understand how this property shows up in a three-dimensional space when we move along a radial trajectory. But the same behaviour of distances from an origin exists here too. It does not mean that at R_o we reach a limit which cannot be crossed, but, at this distance, if we proceed further, our distance from the origin will decrease. It can be rather more clear to say that the gravity of all the matter is strong enough not to allow any material object to leave the region defined by the bulk of matter. Any free particle, moving along a geodesic, will reach eventually the origin of its motion. (Using the analogy of motions of particles in a force field, all particles are forced to move in closed orbits in the field induced by the matter in the universe.)

The case of $k = -1$ is much simpler to comprehend. The denominator of B becomes to be $1 + r^2/R_o^2$, and r can increase up to any value. With r increasing, B becomes smaller, and when $r = R_o$, $B = 1/2$, but r can still increase further. In this model the curvature is negative, there is no limit on distances between points in the universe, and theoretically, a particle which moves along a "straight line" (along a geodesic), can reach infinite distances. This is a model of an open universe. In an open universe gravity is not strong enough to prevent material particles from escaping to infinity. The particles are not bound, and their trajectories are not necessarily closed orbits. We shall see later that the sign of k has a significant meaning in connection with the time evolution of the model. It is important to note that in both cases (in the open model and in the closed one) when $r \ll R_o$, the difference between the two models is very small. Observations in the close vicinity of the observer will not yield differences between the two models, and both of them will seem similar to a flat space. Only in observations over very large distances and when r/R_o becomes a significant quantity, only then can differences between the two models be observed. As R_o is of the order of magnitude of the radius of the universe, it is clear that only observations of inter-galactic distances may yield information connected with the curvature of the universe.

9.6. Length Measurement

Let us study what a distance element in this model is. Let us choose two points P_1, P_2 which lie on the same radial line, and their coordinates are r_1, r_2. The angular differences between these points vanish ($dr = r_2 - r_1$; $d\theta = d\varphi = 0$). The distance element dl is:

$$dl = \frac{R(t)dr}{\sqrt{1 - \frac{kr^2}{R_o^2}}}. \tag{9.8}$$

When we deal with small distances, $r \ll R_o$, the second term under the square root can be ignored, and we obtain:

$$l_{12} = r_{12} \cdot R(t) \tag{9.9}$$

where l_{12} is the distance between the two points, and $r_{12} = r_2 - r_1$. Equation 9.9 is equivalent to the distance in a flat space, in which $k = 0$. Thus for short distances, the distance between two points is proportional to the coordinate difference between them. Still, we have the scaling of the time dependent factor $R(t)$. When we deal with large distances, the second term under the square root cannot be neglected, and we have to integrate equation 9.8:

$$l_{12} = R(t) \cdot \int \frac{dr}{\sqrt{1 - kr^2/R_o^2}}. \tag{9.10}$$

The integration yields different results for different values of k. When $k = 1$, we find:

$$l_{12} = R(t) \cdot R_o \cdot \arcsin\left(\frac{r}{R_o}\right)$$

The function $\arcsin(x)$ is always larger than x, and hence $l_{12} > R(t) \cdot r_{12}$. The distance is not linearly proportional to the coordinate difference r_{12}. When $k = -1$, the integration of equation 9.10 yields: $l_{12} = R(t) \cdot R_o \cdot \text{arcsinh}(r/R_o)$. The function $\text{arcsinh}(x)$ is always smaller than x, and hence $l_{12} < R(t) \cdot r_{12}$.

9.7. Hubble's Constant

In the two cases we dealt with ($k = 1, -1$), the expression obtained for l_{12} includes a spatial dependent part and a time-dependent part, $R(t)$, and we may write l as a product of two functions, one is spatial dependent, $F(r)$, and the other is $R(t)$. So, $l = F(r) \cdot R(t)$. Let us define an entity, H, as the relative change of distance or the rate of change of distance divided by the distance:

$$H = \frac{\dot{l}}{l} = \frac{F(r) \cdot \dot{R}(t)}{F(r) \cdot R(t)} = \frac{\dot{R}(t)}{R(t)} \tag{9.11}$$

where $\dot{l} = dl/dt$. We find that H depends on time alone, and does not depend on the spatial coordinates. This means that if the universe is expanding, then the relative change of distance is equal at all points in the universe (but not necessarily constant in time). This quantity is called Hubble's constant. According to equation 9.11, two material objects, for which the coordinate distance between them is constant, will seem to be recessing from each other with the velocity v, given by:

$$v = \dot{l} = \dot{F(r) \cdot R(t)} = F(r) \cdot R(t) \cdot \frac{\dot{R}(t)}{R(t)} = l \cdot H. \tag{9.12}$$

Equation 9.12 is called "Hubble's law," because Hubble was the one who observed in 1929 that the galaxies are receding from us. This law asserts that the velocity of the mutual recession between two material objects (galaxies) is proportional to their distance from each other. When we observe the galaxies in the universe around us, we find that on the average all the galaxies are moving away from us with velocities which increase with the distance. The further the galaxy is from us, the greater is its velocity of recession. The velocity of a recessing object is measured by the Doppler shift of its radiation, and this shift does not depend on the distance. Hence we have an accurate measurement for recessing objects at distances which are much larger than those which can be measured by any other measurement. Hubble's law, together with the Doppler shift measured for remote objects, is used today to determine the distance to these objects. The value of Hubble's constant is: $H = 1.8 \times 10^{-18} (\text{cm}/\sec)/\text{cm}$. (The units of H are velocity per distance.) In the units used by astronomers it is: $H = 55(\text{km}/\sec)/\text{Mpc}$. (One Mpc is a million parsecs, where one parsec is approximately three light years).

The velocity measured for some object divided by Hubble's constant yields the distance to the object. For some objects whose calculated velocities were found to be a significant fraction of the velocity of light, their distance from us as determined by Hubble's law is some billions of light years.

Hubble's law puts an upper limit to distances available for observations. If we write a distance in the form $l = v/H$, we find that as there is a upper limit for the velocity of a material object which can be observed by us (the velocity of light, c), so there is a maximal distance that can be measured according to this law, and it is $l < c/H$. This distance is a horizon for our observations, and we cannot observe objects beyond this horizon. This does not mean that there cannot exist objects beyond this distance. It means only that if such objects exist, they will be unobservable by us.[2] It should be understood that due to the assumption about homogeneity of the universe, any point may serve as the origin for the coordinate system. Thus an observer who lives beyond the horizon for us, might observe quite a regular universe around him. He will find, from his position, that distant galaxies are moving away from him, with velocities which increase with the distance. For him, our galaxy is unobservable as it is located beyond his horizon.

As Hubble's constant is not constant in time, but (as we shall see later) is decreasing with time, our horizon is expanding, and galaxies which yesterday were beyond the horizon for us, may appear to us tomorrow. In the first epoch of the evolution of the universe, H was much larger than it is today, and the horizon was much closer than it is today. Later on we shall discuss the consequences which emerge from this fact.

9.8.   Time Evolution

Let us study the time evolution of the model. This is done through the equation of $R(t)$. From the field equations of the isotropic coordinate system, we obtain an equation which connects $R(t)$ to the density ρ:

$$\frac{8\pi G}{3c^2}\rho R^2(t) = \left(\frac{R(t)}{c}\right)^2 + k \tag{9.13}$$

where k is the same constant that appeared in equation 9.10. There is another equation which includes both density and pressure in addition to $R(t)$. The equation is:

$$\frac{\rho}{\rho + \frac{P}{c^2}} + \frac{3R(t)}{R(t)} = 0. \tag{9.14}$$

In order to treat two equations which include three variables, we need a third equation. There is another equation which connects the pressure and density, the equation of state, mentioned in Chapter 8. The knowledge of the equation of state at given conditions enables us to solve the equations for three variables, $R(t)$, ρ, P. Which equation of state shall we choose? In the universe today, most of the mass-energy density is contributed by the rest mass of the matter, and the kinetic energy in the universe is negligible. The pressure connected with the kinetic energy of the particles is negligible too, relative to the energy density of the rest mass, ρc^2. Hence, solving the equations for the universe in its state today, we can neglect the pressure and write $P = 0$. (This is a kind of an equation of state which was needed to complete the system of equations.) It should be noted that at the first epoch of the universe, such an assumption is not necessarily true. With this assumption $(P = 0)$, we obtain from equation 9.14:

$$\frac{\rho}{\rho} + \frac{3R(t)}{R(t)} = 0 \tag{9.14a}$$

whose solution is:

$$\rho \cdot R^3(t) = \text{Constant} = \frac{3}{4\pi} M. \tag{9.15}$$

M is a constant of integration, and the coefficient $3/4\pi$ is introduced to fit better into the equation we shall use later. The term $R^3(t)$ is proportional to the "volume" of the universe, and we can assume that M (a product of the "volume" by the density) is proportional to the total mass in the universe. Let us substitute equation 9.15 into equation 9.13 to obtain:

$$\frac{R(t)}{c} = \sqrt{\frac{2GM}{R(t)c^2} - k} \tag{9.16}$$

where k is the curvature constant. The role of k in this equation is crucial, the same as it was in the equation of $B(r)$.

When $k = 0$, equation 9.16 can be solved straightaway yielding:

$$R(t) \propto t^{2/3}. \tag{9.17}$$

From this equation we see that $R(t)$ increases continually with time, but with a decreasing rate. From equation 9.16 we can observe that if $k = 0$ and $R(t) \to \infty$, $R(t)$ tends to zero. Of course $R(t)$ will always increase, but with decreasing rate, and in the far future, this rate will tend to zero. The scale factor of the universe is $R(t)$, and this behaviour represents the time evolution of the universe.

When k is different from zero, the solution of equation 9.16 is more complicated, and the equation may be solved by using a parameter τ by substituting $d\tau = dt/R(t)$. But analyzing equation 9.16 gives an insight into the behaviour of $R(t)$, even without solving the equation.

The case $k = 1$. The left term under the square root should be larger than (or equal to) unity, to prevent the term under the square root from becoming negative. When $R(t)$ increases, the term under the square root decreases, and $\dot{R}(t)$ decreases too, which means that the rate of expansion of the universe decreases. When the left term under the square root equals unity ($2GM/R(t)c^2 = 1$), $\dot{R}(t) = 0$, which means that the expansion of the universe is halted. We know that there is no static solution to such a model, and hence, after the expansion stops the universe will start contracting. Equation 9.13 was obtained for $\dot{R}^2(t)$, and hence it is also valid for contraction, for negative $\dot{R}(t)$. In our earlier discussion we chose a positive sign for $\dot{R}(t)$ because, according to the observations today, the universe is now in a state of expansion. But when $R(t)$ reaches a certain maximal value, the expansion will stop and contraction will start. From the equation we see that the rate of the contraction will increase as $R(t)$ becomes smaller. The contraction will be a symmetric process to the expansion which preceded it.

The case $k = -1$. When k is negative, both terms under the square root in equation 9.16 are positive, and there is no limit to the increase of $R(t)$, which means that there is no limit to the expansion of the universe, as the right wing of the equation never vanishes. Note that the right wing of equation 9.16 is always larger than unity, hence $\dot{R}(t) > c$. The left term under the square root tends to zero when $R(t)$ becomes very large, and the rate of expansion $\dot{R}(t)/c$ approaches unity. The universe will expand to infinity, and the process will not be reversible.

The case $k = 0$. In case that k vanishes, the right wing of equation 9.16 tends asymptotically to zero for very large $R(t)$. The universe will expand to infinity, with a decreasing rate, until the rate tends to zero.

In all three cases we have studied, we have found that $R(t)$ increases with time and $\dot{R}(t)$ decreases with time. Hence, Hubble's constant is decreasing as long as the universe is expanding. This means that this "constant" is equal for all points in space, but it changes with time. For all values of k, the universe expands with the scale factor $R(t)$. As $R(t)$ appears in the denominator of the left term under the square root in equation 9.16, the rate of expansion is decreasing. The limit for expansion depends on the sign of k : $k > 0$ yields a limit to the expansion, while $k < 0$ enables infinite expansion. Thus we find that k determines the character of the universe expansion, in addition to its role in determining the character of the spatial curvature.

Positive k yields positive curvature, which means a universe which is spatially closed and with limited expansion followed by a contraction. The universe is closed both spatially (maximal distance that can be reached at each moment) and temporally (maximal radius of expansion). Negative k yields negative curvature, i.e., a spatially opened universe with unlimited expansion.

9.9.   The Critical Density

The behaviour of the universe—decelerated expansion—reflects the influence of gravity on the expanding bulk of matter. Continuation backward in time of the present state leads to the conclusion that the universe started from a very dense state by a very fast, explosion-like expansion, whose rate decreased until it reached the present rate of expansion. This idea led to the creation of the "big bang" model. According to this model, the universe had a very small volume, and from this state it began with a fast expansion. At the beginning, the character of the expansion was explosion-like, but later, the rate of the expansion was decelerated by gravity. The fate of the universe, whether it will expand forever or recontract, depends on the ratio between the kinetic energy of the matter and the gravitational energy. The sign of k reflects this ratio. If the kinetic energy dominates, k is negative, the deceleration is not sufficient to halt the expansion, and it will proceed forever. If gravity dominates, k is positive, the kinetic energy will drop to zero when the expansion stops, and the universe will contract. Hence, it is important to find out what is the sign of k.

Equation 9.13 connects k, ρ, and $R(t)$. When we substitute $k = 0$ in this equation, we obtain the density as function of $R(t)/R(t)$. Let us denote this density the "critical density," ρ_c. This density suits $k = 0$, and it is the limiting value which separates negative values of k from positive ones. If the density in the universe is higher than ρ_c, then k is positive, and the universe is closed. A density lower than ρ_c shows that k is negative and the universe is open. As Hubble's constant is a time-dependent quantity (which decreases with time), ρ_c is time-dependent too, and it decreases with time. Hubble's constant is known from observations. Hence, measuring the value of the density in the universe today may yield the sign of k. The value of the critical density today is approximately: $\rho_c = 5 \times 10^{-30} \mathrm{gr/cm^3}$. A higher value for the density in the universe will "close" the universe. A lower value for the density will show that the universe is open.

The value of the density in the universe obtained by observations is not reliable enough. We can directly observe "visible mass" (the mass included in stars and galaxies). The visible mass yields a density of $2 \times 10^{-31} \text{gr/cm}^3$, which is smaller by a factor of twenty-five than the critical density. But there is indirect evidence, based on considerations of the dynamics of astronomical objects and other considerations, that there should exist a large amount of "missing mass," a mass which cannot be observed directly (like cold matter which does not emit light); such mass might change the average density significantly, bringing it to a value above the critical density. Intensive efforts are currently being made to observe and measure this missing mass, to determine whether it would suffice to close the universe or not. From general considerations connected to the evolutionary path of the universe in its first epoch, many scientists assume that the average density in the universe is very close to the critical density.

9.10. Red Shift of Light from Remote Galaxies

We noted earlier that due to the expansion of the universe, objects whose coordinate difference between them is kept constant are actually recessing from each other. Let us present the calculation of the red shift caused by this recession, as calculated in a co-moving system. Let us locate the observer at the origin of the coordinate system. Any light source is located then on a radial line, emerging from the origin. The line interval for the light which reaches the origin is:

$$dS^2 = c^2 dt^2 - R^2(t) \frac{dr^2}{1 - \frac{kr^2}{R_o^2}} = 0. \tag{9.18}$$

This equation may be rearranged to obtain:

$$\frac{cdt}{R(t)} = \frac{dr}{\sqrt{1 - \frac{kr^2}{R_o^2}}}. \tag{9.19}$$

The right wing of the equation depends on r only and is constant in time, while the left wing of the equation depends on t only, and it should be constant too. Integrating this equation yields:

$$\int_{t_1}^{t_2} \frac{cdt}{R(t)} = F(r) \qquad (9.19a)$$

where $F(r)$ is a constant which represents the spatial dependent part of the distance between the two endpoints of the interval. Suppose that a light pulse whose time length is Δt_1 is emitted by a source at the moment $t = t_1$. This pulse will be observed by the observer at the origin at the moment $t = t_2$, and its measured time length by the observer is Δt_2. Let us perform the integral of equation 9.19.a twice: first we perform it from t_1 to t_2 and obtain the entity $F(r)_1$, and second we perform it from $t_1 + \Delta t_1$ up to $t_2 + \Delta t_2$, to obtain $F(r)_2$. Since the right wing of the equation is constant, $F(r)_1 = F(r)_2$. Both were calculated along a coordinate difference of the trajectory of the light. $F(r)_1$ was calculated between the beginning of the pulse at the source and the beginning of the pulse at the observer's location. $F(r)_2$ was calculated between the ends of the same two pulses. An integral is an additive quantity and can be divided into parts. Let us divide $F(r)_1$, $F(r)_2$ in the following way:

$$F(r)_1 = \int_{t_1}^{t_1+\Delta t_1} \frac{cdt}{R(t)} + \int_{t_1+\Delta t_1}^{t_2} \frac{cdt}{R(t)} \qquad (9.20)$$

$$F(r)_2 = \int_{t_1+\Delta t_1}^{t_2} \frac{cdt}{R(t)} + \int_{t_2}^{t_2+\Delta t_2} \frac{cdt}{R(t)} \qquad (9.21)$$

By subtracting $F(r)_1$ from $F(r)_2$, we find:

$$F(r)_2 - F(r)_1 = \int_{t_2}^{t_2+\Delta t_2} \frac{cdt}{R(t)} - \int_{t_1}^{t_1+\Delta t_1} \frac{cdt}{R(t)} = 0. \qquad (9.22)$$

We find that the integral of the quantity $cdt/R(t)$ over the duration of the pulse, performed once at the observer location and the other at the source, are equal. If the duration of the pulse is very short relative to the time needed for light to cross the distance from the source to the observer, we may assume that $R(t)$ remained constant along the duartion of the pulse, and we write the integrals:

$$\frac{c\Delta t_1}{R(t_1)} = \frac{c\Delta t_2}{R(t_2)}. \tag{9.23}$$

Let us assume that the length of a pulse equals one wavelength: $\lambda_1 = c\Delta t_1$; $\lambda_2 = c\Delta t_2$. Hence:

$$\frac{\lambda_1}{R(t_1)} = \frac{\lambda_2}{R(t_2)}$$

or

$$\frac{\lambda_1}{\lambda_2} = \frac{R(t_1)}{R(t_2)}. \tag{9.24}$$

The ratio between the wavelengths measured by the source and the observer is equal to the ratio between $R(t)$ at the moment of emitting and $R(t)$ at the moment of receiving. This means that the change in the wavelength is proportional to the expansion of the universe during the time the light was on its way from the source to the observer. Since we know the rate of the expansion, equation 9.24 can be used for distance measurement. The change in the wavelength tells us by what factor the radius of the universe increased since the light was emitted, and from this we can calculate how long the light spent on its way from the source to the observer. This time yields the distance to the source given in light-seconds. It should be remembered that the distance calculated in this way is the distance to the source at the moment the pulse was emitted. Since then the source has probably moved further away due the expansion of the universe. Because of this expansion, the picture of the universe obtained in this way is not the present map of the universe, but a contracted map whose contraction is not uniform. The further the object is from us, the larger the deformation of the picture is relative to the correct map. Hence the picture we obtain is similar to the correct map for the nearby regions; as we look to further distances in the universe, the picture is more contracted.

Comments

1. Einstein's equations are also called the field equations. We shall not present the equations in full, nor will we try to solve them in detail. But we shall use them whenever it might help to make the picture more transparent.

2. We do not "measure" velocities of galaxies and other distant objects. We measure wavelength shifts in the light we receive from them. The quantity used in this connection is Z. ($Z = \Delta\lambda/\lambda$), which is the difference in wavelength between the absorbed and emitted light, divided by the original wavelength. At low velocities, Z is approximately proportional to the velocity of the recessing source (in this approximation $Z = v/c$). Originally, Hubble's law was formulated according to the ratio between the distances to the galaxies and the value of Z measured in their light. At low velocities, the ratio between Z and the velocity of the source is linear. But for high velocities, this ratio should be calculated according to the relativistic formulae, which are very non-linear. The largest Z measured to date was found in light emitted by quasars, and it was close to $Z = 4$.

CHAPTER TEN

Relativistic Astrophysics Phenomena

In this chapter we present several examples from astronomy connected with relativistic effects.

10.1. Quasars

In this connection we should mention the "quasars" (acronym for "quasi stellar radio sources"). These objects, whose main radiation is in the radio wave range, were found to have a very high red shift in their radiation (the shift was found in the visible light of the quasi stellar object). The shift is so high that at first observers could hardly identify the spectral lines. When finally the spectral lines were identified, the distance calculated by these lines was found to be of the order of a few billion light years, which are large distances even on astronomical scale. We know that the intensity of the energy radiated decreases with distance (by a factor of $1/r^2$). Knowing the distance to these sources and measuring the radiation intensity which reaches us enabled scientists to calculate the amount of energy radiated by the source. The calculations yielded enormous numbers. The amount of energy radiated by such an object was found to be a few orders of magnitudes larger than the energy radiated by the entire Galaxy. On the other hand, calculations connected with the rates of change in that radiation led to the conclusion that these objects are much smaller than a typical galaxy. There was no satisfactory explanation for the existence of such an intense source. Some astronomers argued that the distance to these objects is much smaller than calculated and that these objects are actually located in our Galaxy. This assumption solved the problem of the high intensity of the source, but it required an explanation for the high red shift observed.

We already know about another source for red shift; gravity. Thus for several years there was a controversy about the source of the red shift of the quasars. The disagreement was between the "cosmological red shift" supporters, who argued that the red shift is caused by the large distance of these objects from us, and the "local red shift" supporters who held that these objects are located in our galaxy and the source for their red shift is a high gravitational field existing in and around them. The weakness of the approach of the local red shift is that, if quasars are a local phenomena in our galaxy, we should expect that their distribution in space will somehow suit the structure of the galaxy. Such coordination was not found. During the last few years many more quasars have been discovered, and to date a few hundred of them are known. In the light of some of them, the phenomenon of the gravitational lense was found. Such phenomena usually occur over a very large range of distances, and it is impossible that it could be found in a light from a source located in our galaxy.

Today, most astronomers believe that the red shift in the light emitted by quasars is cosmological. They are very remote objects, and we still have no satisfactory explanation for the enormous amount of energy which they radiate. All the quasars were found to be very far from us, and the radiation we observe was emitted billions of years ago. If the quasars were regular phenomena of stellar evolution, we could expect to find some of them also in our close vicinity, which means at a later epoch of the evolution of the universe. Since we do not find any quasars in our vicinity, we conclude that the existence of the quasars characterizes early stages of the evolution of the universe. When we possess a better understanding of the nature of the quasars, we shall, perhaps, better understand the course of the evolution of the universe. Recently, other objects were discovered which are similar to quasars, but their range of radiation wavelength is much broader, and not especially in the radio range. Hence, today we talk not about quasars, but about quasi stellar objects (QSO).

10.2. High Density Matter

As already mentioned in Chapter 7, relativistic effects become important when matter density is high enough to create non-linear effects in the gravitational energy. A single atom will not create relativistic effects. Schwarzschild's radius of a hydrogen atom is of the order of 10^{-55}cm, and from experiments we know that the radius of the nucleus is of the order of 10^{-14}cm. Thus such a nucleus is far from creating relativistic effects. If we construct an object by a very dense packing of protons, the mass of the object will increase as the third power of the radius of the object, r. Hence the quantity $2GM/c^2r$ which is the measure for the relativistic character of the object increases as $r^3/r = r^2$. Thus an object consisting of dense packed protons with a radius of a few kilometers will form a black hole. But we know that protons are not bare in nature. They are surrounded by electrons bounded by electric forces. These electrons, whose mass is negligible, form with the nucleus an atom whose radius is ten thousands times the radius of the nucleus. Hence, matter constructed of atoms, even in the most dense packing, has a much lower density than the density of matter constructed of nuclei. If we want to create a black hole from regular matter—atomic matter, with density of one gr/cm^3—we have to construct an object whose mass is about 10^9 solar masses. We are not aware of the existence of a single object consisting of such vast amounts of mass.

The masses of most stars known to date are of the order of magnitude of one solar mass. Objects of this order of magnitude could turn into black holes only if they were packed densely. In order to form such a close packing, one has to overcome the electric forces which prevent mutual penetration of neighbouring atoms. The interaction between the atoms is represented by the pressure created in the matter. The density is determined by the balance between the gravitational force and the gradient of the pressure. A material object stays in an equilibrium when the gravity, which acts towards collapse, is balanced by the force created by the pressure. A relatively small mass, of the order of magnitude of the earth, does not create a large gravitational force, and the inter-atomic forces (formed in solid matter) are sufficient to form the equilibrium, known to us as the solid material which compose the upper layer of the earth. In objects with higher masses (of the order of magnitude of the solar mass) the inter-atomic forces of solid matter cannot balance gravity, and the object contracts until the high temperatures formed at its center create high pressure which balances the gravitation.   The nuclear reactions which take place at these high temperatures release energy, which replaces the energy lost from the star by radiation.   As long as there is enough nuclear fuel to supply the energy lost by radiation, the star will stay in a state of equilibrium.   The characteristic lifetime of a star of the order of magnitude of the sun at this evolutionary phase is about ten billions years.   When the nuclear fuel at the stellar core is consumed, the star starts cooling, the pressure drops and the star contracts.   When the matter reaches a density of the order of $10^5 \text{gr}/\text{cm}^3$, a pressure is formed which is actually independent of temperature.   This state is called a "degenerate state," and the degeneracy pressure can stop the contraction, even when the temperature drops to zero in Kelvin scale. Such stars are called "white dwarfs."   They do not produce nuclear energy. The radius of such a star which has a mass of the order of magnitude of the sun is about the radius of the earth.   The equations for the degenerate state were calculated by Chandrasekhar in 1942.   From these equations it is found that in cold stars, whose masses are lower than 1.4 solar mass, the degeneracy pressure can hold against gravitational collapse.   This mass is called Chandrasekhar's mass.   With masses higher than this limit, gravity will overcome the degeneracy pressure, and the star will contract further.   Until recently, it was believed that Chandrasekhar's mass was the upper limit for stable objects, and that heavier objects must collapse to a black hole.

10.3.   Neutron Stars and Black Holes

In recent decades it was found that matter can form a denser stable state than the degenerate state, a state which can resist gravitational collapse. This matter is called neutronic matter. How does such matter form?

When a degenerate matter collapses, the electrons are compressed into the atomic nucleii.   They combine with the protons to form neutrons, and a neutron-rich matter is formed.   When all protons are converted into neutrons, a neutronic matter is formed.   The equation of state of this matter shows that it can develop very high pressure.   The characteristic density of the neutronic matter is about $10^{14} \text{gr}/\text{cm}^3$.   Neutronic matter can support the star against gravitational collapse, even for masses higher than Chandrasekhar's limit.   The equation of state of this matter is not known yet for sure; consequently the upper mass limit for a stable neutron star is not known either. It is believed that the limit is somewhere between 2–4 solar masses. Neutron stars with higher masses than these values will continue to collapse into denser configurations, and they might turn into black holes.

Many white dwarfs are known, and it is assumed that they form the last stage of stellar evolution for stars of their masses. The masses of most of the white dwarfs are around the value of 0.55 solar mass, and few approach the solar mass. Recently, numerous very dense stars were observed and they were identified as neutron stars. The identification is based on the mass of these stars. From their radius calculated from observation, their density was calculated, and found to be around $10^{14} gr/cm^3$. Many of them were found in or very close to remnants of super-nova explosions, and it is believed that they were formed during such explosions. The neutron stars rotate very fast—up to a few hundreds rotations per second. They radiate pulses whose frequency is equal to the rotation frequency. When they were first discovered in the 1960's their nature was not clear, and the only clear fact was their pulse radiation. Hence, they were called pulsars. They lose energy by radiation, and their rotation is slowed down. We may assume that those that we can observe are those whose rotation velocity is still high and whose pulse radiation is strong.

We cannot observe a black hole directly because such an object, by definition, emits no radiation. Finding and identifying a black hole is only possible by indirect observations. Such a possibility exists when a black hole is found as a companion of a visible star. In a binary system, the motion of each one of the components depends on the total mass of the system. Sometimes it is found that a star performs such motions which show clearly that it is influenced by a companion. Such a system is called "astrometric binary system." Many binary systems were observed in this way, and in part of them, the companion was found later. By observing the motions of the visible part of the system, the mass ratio of the two stars may be calculated. If we have a good estimate of the mass of the observed companion, we can estimate the mass of the unobserved one.

One of the serious candidates to be a black hole is a star known as Cygnus X-1. (This name means that this star, or its companion, is an x-ray source in the constellation Cygnus (Swan). The number 1 indicates that it is the first x-ray source discovered in Cygnus.) The calculations for the unobserved companion lead to a possible mass of ten times the solar mass. Since the unobserved star is a cold one (no visible radiation), it cannot create thermal pressure to support the gravity of such a big mass against collapse. Thus, researchers are tempted to believe that this unobserved companion is a black hole.

Another possibility for identification of a black hole is through the behaviour of the matter attracted towards it. A black hole has a very strong gravitational field around it, and the close region to Schwarzschild's radius is strongly affected. Loose matter in the vicinity of the black hole, like inter-stellar gas or matter ejected from neighbouring star, will be attracted and accelerated towards the black hole. Once the matter is absorbed into the black hole, no sign of it will be seen any more. But we can observe that matter when it is on its way towards the black hole. When the distance between two companions of a binary system is very small, the gravitational interaction between them is very strong. If one of the companions is a dense star (white dwarf, neutron star or a black hole) and the other is a regular star or a red giant star with an inflated envelope which is not strongly bounded, the second star will lose mass which will be accreted onto the dense star. Since the system is rotating, each particle has a certain amount of angular momentum. The material flowing towards the dense star will not flow in a head-on trajectory, but will approach the dense star in a spiralic motion, creating an accretion disc rotating around the dense star. Due to the interactions between the particles in the accretion disc, they will gradually lose energy and angular momentum and come closer and closer to Schwarzschild's radius, until they are absorbed. This process may last for a long time, and the energy lost by the particles is radiated in the form of high energy photons—usually x-radiation. In such a complicated system, several types of radiation are created in different regions of the system. Each type of radiation may have periodical changes with characteristic frequency. Figure 10.1 may demonstrate such a system which includes an accretion disc.

By analyzing the radiation with the frequency characterizing each type of radiation and dynamical considerations of the system and the behaviour of the visible components of the system, researchers try to construct models of the system. In such a model, the mass of the accreting star has a crucial role. If it is necessary to include a very massive dense star in the system, it seems that this star is a black hole. The presence of an x-ray source in Cygnus X-1 strengthens the suggestion of the existence of a black hole in this system, but it is still not well established.

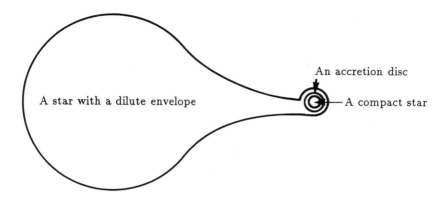

FIGURE 10.1. An accretion disc around a compact star. A close binary system of a regular star and a compact star (e.g. white dwarf). The star with the dilute envelope loses mass to the compact one. The mass transferred has angular momentum and creates an accretion disc around the compact star.

10.4.   Gravitational Waves

It is most convenient to study the phenomenon of gravitational waves through the linear approximation to Einstein's equations.

As an example of a linear treatment, let us start with the "almost Lorentzian" coordinate system. When no gravity is present, the space is flat. If a weak gravitational field exists, we can treat it as a small perturbation over the flat space. The flat space is characterized by the Lorentzian system (of the STR), and the system with weak perturbation is an almost Lorentzian system. In such a case we can write the metric tensor in the form:

$$\mathbf{A} = \eta + \mathbf{h} \tag{10.1}$$

where η is the Lorentzian (STR) metric tensor, and the absolute value of each component of $\mathbf{h}$ is very small. $|h_{ij}| \ll 1$. The Lorentz transformation, as given in Chapter 5 is:

$$\Lambda = \begin{pmatrix} \gamma & 0 & 0 & -\gamma\frac{v}{c} \\ 0 & 1 & 0 & 0 \\ 0 & 0 & 1 & 0 \\ -\gamma\frac{v}{c} & 0 & 0 & \gamma \end{pmatrix}. \tag{10.2}$$

If we operate with this operator on **A** as given in eq. 10.1 we find:

$$\Lambda \cdot \mathbf{A} \cdot \Lambda = \Lambda \cdot \eta \cdot \Lambda + \Lambda \cdot \mathbf{h} \cdot \Lambda \tag{10.3}$$

and as $\Lambda \cdot \eta \cdot \Lambda = \eta$, we obtain:

$$\Lambda \cdot \mathbf{A} \cdot \Lambda = \eta + \bar{\mathbf{h}} \tag{10.4}$$

where $\bar{\mathbf{h}} = \Lambda \cdot \mathbf{h} \cdot \Lambda$. This means that after the transformation, the almost Lorentzian form is preserved. We can treat the space as a background flat space on which a curvature "tensor" **h** is defined. (The name tensor is written here "between apostrophes" as **h** is not a real tensor, and it does not have the mathematical properties required of a tensor.) Any physical field existing in this space can be described as a function of h_{ij}.

Consider a region in which the gravitational field is very weak, but not static. Such a region may be one that is far away from a strong gravitational source, with fast changes occurring in it. Due to the distance from the source, the field it induces is weak, but the changes occurring at the source occur at the distant field too, in a time delay needed for the change to cross the distance. Such a region may be treated by the weak field approximation.

As might be remembered, the left wing of Einstein's equations includes second derivatives of the components of the metric tensor and products of the first derivatives of these components. The components of η are constants and their derivatives vanish. Hence, the derivatives of A_{ij} include only the derivatives of h_{ij}. The components of **h** are very small, and so are their derivatives. Hence, the products of the first derivatives are products of very small terms, and in the linear approximation we are using now, they can be ignored. Thus the left wing of the equations includes second derivatives of h_{ij} only. By an appropriate choice of the coordinate system, we can find a state in which mixed second derivatives vanish and we are left only with terms which include non-mixed second derivatives (second derivative with respect to x, second derivative with respect to y, second derivative with respect to z, and second derivative with respect to t). The left wing of Einstein's equations can be written:

$$-\left(\frac{\partial^2}{\partial x^2}h + \frac{\partial^2}{\partial y^2}h + \frac{\partial^2}{\partial z^2}h\right) + \frac{1}{c^2}\frac{\partial^2}{\partial t^2}h = \left(\frac{i}{c^2}\frac{\partial^2}{\partial t^2} - \nabla^2\right)h \quad (10.5)$$

where ∇^2 is an operator which represents a sum of second derivatives with respect to the spatial coordinates. Since we are at a region which is very far from the gravitational source, the energy-momentum tensor vanishes, and the right wing of the equations is nil. The form of the equations will be:

$$\left(\frac{1}{c^2}\frac{\partial^2}{\partial t^2} - \nabla^2\right)h_{ij} = 0. \quad (10.6)$$

The form of this equation is known in different branches of physics, and it is a wave equation for a region which does not include the wave source (in a region which includes the wave source, there will appear on the right wing of the equation a term which represents the source). The conclusion of this equation is that the components of the "tensor" h have a character of a wave motion. These waves are called gravitational waves. Analyzing the equation and its solutions (which we shall not present here) shows that the gravitational waves proceed with light velocity, c, and the direction of the oscillations of these waves is perpendicular to the direction of motion. In these properties they are similar to electromagnetic waves.

How do these waves influence the space they are passing through? This we may study by analyzing the metrics we have used. We defined the metric tensor as a sum of the STR tensor and h, as a "tensor" of small perturbations over the flat space. Wave behaviour is characterized by the fact that, if we choose a fixed point in space, we shall find that the function which represents the wave (h in our case) performs periodical motions. The concept of a wave means that at each point in space there are small oscillations of the curvature around the state of a flat space, which is the state of equilibrium of the system. If we observe the whole space, we see that these periodical disturbances proceed over the flat space with the velocity c.

Up to here we described the gravitational waves by the treatment of the linear approximation of Einstein's equations. The wave equations may be obtained by the use of the full version of Einstein's equations without approximations.[1] But the treatment will be complicated and tedious, and it will be more difficult to observe the physical contents of the results. Hence, usually the calculations concerning gravitational waves are based on the linear approximation of the equations.

10.5.   How Gravitational Waves are Created

Radiation of any kind may be created only where accelerations exist. An acceleration of electric charges creates electromagnetic radiation, and acceleration of masses can create gravitational radiation. A system in which periodical motions exist, also has accelerations, and hence, systems which include periodical motions are usually sources for radiation. In a source of electromagnetic radiation, the main source of the radiation is translational motions of charges. (In emitting antenna of radio and TV, the charges are accelerated in a periodical translational motion along the antenna.)

A periodical translational motion of charges is called an oscillating dipole, and the radiation emitted due to such motion is called a dipole radiation. Rotational motions also include accelerations and can create radiation, but the amount of radiation emitted in rotational acceleration is much lower than that which is emitted by dipole radiation. The radiation created by rotational accelerations is called quadrupole radiation. (There are other kinds of motions which create quadrupole radiation.) When both dipole and quadrupole radiation are created in a source, the first will dominate, and the quadrupole radiation will be hardly observed. When there is no dipole radiation, however, the quadrupole radiation will be observed clearly, but its intensity will be very low.

One main difference between electromagnetic and gravitational radiation is that in the later there is no dipole radiation. The reason for this is that dipole radiation is proportional to the (squared) product of the charge by its acceleration. In a system which includes several charges, we have to sum the products of all the charges by their accelerations. In gravitation, the relevant charge is the particle mass, and the product of the mass of a particle by its acceleration is the force. The law of momentum conservation (or the law of action and reaction) states that in a closed system the sum of all the forces should vanish. Hence the sum of the term which yields dipole gravitational radiation vanishes, and the dominant radiation will be the quadrupole radiation. This fact should be considered when planning detectors for gravitational radiation. For the same reason the gravitational radiation is very weak, and there are many difficulties in observing this radiation. In a spherical symmetric system, any radial motion has the character of dipole motion. Hence gravitational effects which are spherical symmetric (like collapse into a black hole, pulsation of a star) will not give rise to gravitational radiation. On the other hand, gravitational processes which include non-radial accelerations will create gravitational radiation. Hence, potential sources for gravitational radiation include a collapse which occurs through rotation and binary systems with high frequency periods.

Through the 1960's, detecting systems for gravitational radiation were developed. The pioneer in this field was J. Weber (1964) from the University of Maryland, and he was followed by other groups from the United States and Germany. Weber's[2] detector is based on a big metal bulk suspended in such a way that it will be able to absorb any oscillation and respond to it. In his experiment he used a cylinder of aluminium with a mass of 1.5 tons, which was suspended in a vacuum. The cylinder is supposed to start vibrating due to periodical changes in the curvature of space caused by the gravitational waves. Measuring the amplitude of the oscillations in the cylinder enables us to calculate the magnitude of the gravitational radiation which gave rise to the oscillations. Theoretical calculations of the expected magnitude of the waves showed that it should be very low. Although a star with high rotational frequency loses a lot of energy by gravitational radiation, the intensity of the radiation expected to be received on earth is less than one Watt. This intensity is hardly above the standard noise caused by the thermal motions of the molecules in the detector. Further, the observer should distinguish between disturbances caused by local sources (earthquakes, motion of cars, airplanes etc.) and signals approaching from far away sources. In order to isolate signals which approach from remote sources, two identical detectors were constructed, and they were located in two places with thousands of kilometers between them. Each detector detected many signals, but only signals which were recorded simultaneously by both detectors were considered as gravitational waves arriving from remote sources.

A system with suitable characteristics to be a source for gravitational radiation is a system which includes pulsars, (very dense stars rotating with very high frequencies). Among the first pulsars observed were some that were rotating at a rate of thirty cycles per second. Recently (1982) a pulsar which rotates with 642 cycles per second was discovered. (In such a star, the linear velocity of the rotation on its equator is about 0.13 of light velocity.) A year later, a pulsar which rotate with 164 cycles per second was discovered. It is suggested that the last two were formed from merging a pair of small stars which were very close to each other, and they lost mass while mass was transferred between them. Because of the fast rotation of the pulsars, strong centrifugal forces are formed which cause deformation of the star which turns into an ellipsoid with high eccentricity. Oscillations in an ellipsoidical object have quadrupole moment. This quadrupole moment, plus the fast rotation, may cause relatively strong gravitational radiation. Such systems were subject to the observations of Weber and others, but to date, no evidence has been found to prove a positive absorption of gravitational radiation.

There is indirect evidence for the existence of gravitational radiation. A pulsar emits a very sharp pulse of radiation (this is the source for its name), and its frequency of rotation can be measured very precisely. It is found that the frequency of rotation of the pulsars is decreasing. When we calculate the energy losses which fits the decrease in the frequency, it is found to be of the order of magnitude of the energy that such a star should radiate by gravitational radiation. This means that the pulsar radiates due to its fast rotation. This radiation is actually an energy loss, and the decrease of the energy of the star causes a decrease of the frequency of rotation. From such measurements we obtain indirect evidence that the star indeed radiates in the form of gravitational radiation.

References

1. Misner, C. W., Thorne, K. S., Wheeler, J. A., *Gravitation.* Freeman and Co., San Francisco, 1973.

2. Chiu, H. Y., Hoffman, W. F., *Gravitation and Relativity.* (J. Weber in) Benjamin Inc., New York, 1964.

CHAPTER ELEVEN

Epilogue

The solutions of Einstein's equations given in the preceding chapters are the most simple of their kind. Some of them are typical solutions for the description of the curvature created around an isolated object (exterior solutions), while the others deal with the entire universe as a unit (interior solutions).

There are numerous other solutions, more sophisticated and more complicated than those mentioned above, attempting to cover the numerous varieties which exist in nature, but the treatment of these solutions requires higher knowledge of mathematical techniques, and they will not be treated here. We can mention the interior solutions for isolated objects, where use is made of detailed equations of state. Using these equations of state, the structure of very dense objects, such as neutron stars, is calculated. In these dense objects, the weight of the relativistic effects is significant. Equations of state for "quark matter" have been developed. It is suggested that this matter is more dense than neutronic matter, and with the help of these equations, the structure of objects which collapsed beyond the state of neutronic matter may be studied. Using other types of interior solutions, people are trying to study processes which occur inside objects that are in a state of collapse to a black hole. There exist exterior solutions of isolated objects which have non-spherical symmetry, like rotating black holes. The rotation creates interesting phenomena in the black hole, some of which are observable. Thus there are more chances to observe phenomena connected with a rotating black hole than those connected with a static one.

Cosmological models are studied to trace the evolution of the universe and to modelize the first epoch of the evolution, close to the big bang. Different models for the first few minutes of the existence of the universe result in different physical parameters in the present universe. By matching the detailed cosmological model to the present astronomical observations, we can learn much about the evolution of our universe, and also deduce what is awaiting us in the future. From this matching, one can learn about the behaviour of matter in a state of high energy density, a state that existed in the first epoch of the existense of the universe. The interested reader may find more details in textbooks dealing with these topics, such as those written by Weinberg[1,2], Zeldovitch and Novikov[3], Misner, Thorne and Wheeler[4].

The theory of relativity is general in the way that it unifies an entire class of phenomena connected with gravitation and inertial forces into a geometrical picture. It was natural for Einstein (and others) to believe that such a general theory could be generalized to include other natural phenomena, such as the electromagnetic interactions. Einstein invested a lot of effort in trying to generalize the theory of relativity over electromagnetism, but without success.

At the beginning of this century, scientific knowledge included two main kinds of interactions: gravitational and electromagnetic. If Einstein's hope to include electromagnetism in the theory of relativity had been realized, he would have reached a generalization of all the natural phenomena known at his time in one theory. Since then things have become more complicated. In addition to the gravitational and electromagnetic interactions, a new class of forces was discovered—nuclear forces. The behaviour of these forces is very different from the forces that were known before. While the gravitational and electromagnetic forces are long-range forces and fall with distance like the square of the distance, nuclear forces are very short-range forces. They fall with distance very fast, and the characteristic range for their action is the radius of the nucleus. Later it was found that actually there are two nuclear forces, which were called the "strong force" and the "weak force." Thus, at mid-century, science knew already four classes of forces. (Recently, in 1986, people began talking about a fifth force. This force, if it exists, is supposed to act on the constituents of the nucleus, the protons and the neutrons.)

Although the number of types of forces in nature is greater now, ways were found to unify at least part of them into more general classes. In the 1970's, the "unified theory" was developed, which unified the weak force with the electromagnetic forces, and today they form one class. Now a "Grand Unified Theory" (GUT) is under construction which tries to unify the strong force with the weak and electromagnetic forces. Scientists believe that in the future a super-GUT theory will be found that will unify gravitation with all other forces. The assumption is that at the first moments of the existence of the universe, when a very high energy density existed, all the forces formed one class, and there was no difference in their way of action. Hence, the chances of finding suitable conditions to verify the super-GUT are in the vicinity of very high energy densities, like those which existed at the very beginning of the universe. Maybe the coming generations of particle accelerators will be able to create such high energy densities needed for the creation of the super-GUT. When these plans are realized, the dream of Einstein to find a unified theory will be realized too, although in a way different from what he might have thought.

To conclude our study of the GTR, we may ask two questions: What did we learn from this theory, and for what can it be used? As for the first question, it is my hope that the reader has gained a clear view of the leading idea of the theory, which is that all the phenomena connected with gravity and with inertial forces can be fully represented by the four-dimensional geometry of space. All objects' motions subject to these forces are uniquely determined by the geometry. Thus, the study of the geometry yields a complete and general picture of the nature of the motions of all material objects. This is why the metric tensor plays such a central role in the theory. All the information about the geometry is included in the metric tensor. Hence, the components of the metric tensor are the variables of the theory, and when they are found from the solution of Einstein's equations, they supply all the vital information about the physics included in the GTR.

The answer to the first question already contains part of the answer to the second. The GTR provides a general view of a certain class of phenomena, thus yielding a better understanding of nature. From a "practical" point of view, this theory does not provide us with efficient tools which may improve our standard of living. Nothing like this is expected from the GTR in the future. The use of the GTR will continue (at least for the near future) to be a theoretical one. But this does not reduce its importance. Natural human curiosity, which is never satisfied, and the strong motivation to study and understand the world around us were, and still are, the main drive for the evolution of mankind. From this point of view we cannot exaggerate the contribution of the GTR. We could not imagine our understanding of nature today without the brick added by the theory of relativity of Albert Einstein.

References

1.   Weinberg, S. *Gravitation and Cosmology.* Wiley & Sons, New York, 1972.

2.   Weinberg, S. *The First Three Minutes.* Basic Books Inc., New York, 1977.

3.   Zeldovitch, Y. B., Novikov, I. D. *Relativistic Astrophysics.* University of Chicago Press, Chicago, 1971.

4.   Misner, C. W., Thorne, K. S., Wheeler, J. A. *Gravitation.* Freeman and Company, San Francisco, 1973.

Glossary

Acceleration—The rate of change of the velocity. The acceleration is the derivative of the velocity with respect to time. According to the second law of Newton, the acceleration of an object is proportional to the sum of forces acting on this object, and inversely proportional to its mass.

Astrometric—Measuring stars. An astrometric property of a star is a property identified indirectly, through observations performed on other stars, e.g., an astrometric binary stellar system is a system in which only one of the binary stars is optically observed. The motion of the observed star shows that it is influenced by the presence of an unobserved companion.

Asymptotic—When a curved line approximates a straight line to an almost convergence, we say that the straight line is an asymptote to the curved one. At infinity (i.e., at a very large distance), the straight line is tangent to the curved one and can represent it.

Big circle—The big circle on a spherical surface is a circle that is contained in a plane that passes through the center of the sphere. Any circle on a spherical surface that is not a big circle is shorter than the big circle. The longitudes in the geographical mesh of coordinates are big circles.

Cartesian—After Descartes. A coordinate system whose axes are straight lines.

Causality—The principle of causality states that when a certain event is the cause of another event, then in any coordinate system the causal event will precede the resultant event. The order of events in a causal chain of events is an absolute property. This order will always be kept in any coordinate system chosen to describe the events. The interval between two events connected by a causal order is a time-like interval.

Centripetal—Attraction toward the center. In order to maintain a circular motion, a centripetal force is required to balance the centrifugal force created through the motion.

Centrifugal—Repulsion from the center. This force appears when an object moves in a circular motion. The direction of this force is outward, along the radius of the rotation.

Co-moving—A co-moving coordinate system is a system that is attached to an object and that moves with it. In such a system, the object is always at rest, and only local motions that take place inside the object will be registered in a co-moving system.

Covariant—Vary together. A system of several variables is called a covariant system if all these variables changes in such a way that the geometrical or physical interrelations between the variables are conserved.

Physical equations whose formulation contains only physical quantities and physical relations, and that do not depend upon a choice of a special coordinate system, are called covariant equations. A covariant formulation of natural laws suits the principle of relativity, since it is valid in any coordinate system, without preference to any one system.

Covariant derivative—A derivative calculated in such a way as to also include the change of the variable due to the change in space.

Derivative—The ratio of the change of one variable to the change of another, upon which the first variable depends. In general, this ratio is not constant and it must be calculated at each point. This calculation is performed by finding the ratio between infinitesimal changes of the variables—a ratio between differentials.

Formally we write:

$$y'(x) = \frac{dy}{dx} = \lim_{\Delta x \to 0} \frac{\Delta y}{\Delta x}$$

where $y'(x)$ is the derivative of y with respect to x, Δx is the change in x, dx is the differential of x, and dy/dx is the ratio between the differentials dy and dx. $\lim_{\Delta x \to 0}(\Delta y/\Delta x)$ is the ratio between the changes in y and in x when the magnitude of the change in x tends to zero.

A derivative of a variable x with respect to time yields the rate of change of x, which is the velocity in the x direction. The rate of change of the velocity $v'(t) = dv/dt$ is the acceleration.

When a variable depends upon several different independent variables, we have to calculate its derivatives with respect to each one of the independent variables. Such derivatives are called "partial derivatives," and they are written in the form $\partial y/\partial x$ (instead of dy/dx).

Differential—A very small (infinitesimal) difference. When dealing with flat surfaces or with systems where the changes take place at constant rates, we can deal with any magnitude of the changes of the variables. When dealing with curved surfaces or with systems where the changes take place at varying rates, we have to deal with very small changes of the variables. These small changes are called "differentials." In calculating an integral, we calculate the sum of differential quantities. The quotient of two differentials implies the derivative of the variable that appears in the numerator with respect to the variable that appears in the denominator.

Dipole—An object possessing two opposite poles, such as a magnet possessing a north pole and a south pole, or an isolated rod charged by a positive charge at one end and a negative charge at the other end. When an electric dipole oscillates, there exists a linear acceleration of the charges, and this acceleration creates an electromagnetic radiation.

Doppler effect—The phenomenon of the change of wavelength of a radiation caused by the relative velocity between the emitter and the receiver. When this relative velocity is an approach, the wavelength decreases and the wave frequency increases (blue shifting). When the relative velocity is that of recession, the wavelength increases (red shifting).

When we deal with waves that move in a material medium, such as sound waves in air or in water, there is a difference between the situation in which the emitter is at rest and the receiver is moving relative to the medium, and the situation in which the emitter is moving and the receiver is at rest relative to the medium. This difference between the two situations does not exist for waves moving in vacuum—such as electromagnetic waves and gravitational waves. When the relative velocity between the receiver and the emitter is low relative to the wave velocity, we have $Z = (\Delta\lambda/\lambda) \simeq (v/c)$, where $\Delta\lambda$ is the shift in the wavelength, v is the relative velocity between the emitter and the receiver, and c is the wave velocity. When v/c is not small, the expression for $\Delta\lambda/\lambda$ is more complicated.

Eccentric—Out of center. The eccentricity of a curved line expresses how much this line deviates from symmetry. The eccentricity of an ellipse expresses how much the ellipse deviates from a circle. If a is the semi-major axis of the ellipse and b is its semi-minor axis, then the eccentricity, ε is given by:

$$\varepsilon^2 = 1 - \frac{b^2}{a^2}.$$

Einstein, Albert—(1878–1955). Born in Germany. Began his scientific work at the beginning of the 20th century. During the years 1905–1906 he published three articles, each one of them a breakthrough in a different field: (1) The Special Theory of Relativity; (2) an explanation of the photo-electric effect that was the basis for the quantization of the light phenomena, and (3) an explanation on the specific heat of crystals at low temperatures.

Einstein became famous especially for the Theory of Relativity, which he developed: the Special Theory in 1905, and the General Theory in 1915. Einstein is considered one of the greatest scientists of our times.

Energy—The common definition of energy is "the ability to perform work." This definition does not cover all the forms of energy.

The performance of work is characterized by changes in the energy. The object that performs the work loses energy, while the object on which the work is performed gains energy. Conservation of energy is one of the basic laws of nature. It states that the general energy balance is always kept. When performing relativistic calculations of energy, we also include in the energy balance the rest mass of the matter given by the formula:

$$E = mc^2.$$

Kinetic energy—The energy connected with motion. The kinetic energy of an object is the energy it acquires when brought into a state of motion. A Newtonian calculation of the kinetic energy, E_k, of an object with mass m is given by:

$$E_k = \frac{1}{2} mv^2.$$

Potential energy—The energy contained in a certain state. The potential energy of an object at a certain state is the energy it acquired when brought to this state. When it returns to its original state, it will release the potential energy it gained when brought to the present state.

Equation of state—An equation that defines the relations between the thermodynamic variables in a system. Usually it is used to express the pressure as a function of other thermodynamic variables, such as the density and the temperature.

Equivalence—Two different objects are equivalent concerning a certain topic if their influence, or the way they act concerning this topic, is equal.

Extremal—Points of maxima and minima of a phenomenon are the extremal points of the phenomenon.

Feedback—The influence of the results of an action on the cause of this action.

In a positive feedback, the result of the action causes an enhancement of the cause of that action. This implies the enhancement of the result, which will enhance the cause, and so on and so forth. In a positive feedback there is a mutual reinforcement between the cause and the result, and the process will be infinitely accelerated.

In a negative feedback, the result of the action causes a weakening of the cause of the action. This implies a weakening of the result, which will allow the cause to enhance. An enhancement of the cause will enhance the result, which will weaken the cause. Thus there is a balance between the cause and the result that keeps the process on a certain (constant) level of activity. Any deviation from this balance implies a reaction toward returning to the balanced state. Systems that contain negative feedback mechanism tend to balance themselves in a "steady-state" equilibrium.

Four-dimensional diagram—A diagram that includes the spatial axes and the time axis. Since drawing on a flat paper yields only a two-dimensional diagram, usually only one spatial axis is drawn with the time axis. If the system is aligned in such a way that the motion in the system is along the spatial axis drawn in the diagram, the diagram includes all the information needed for the description of the motion.

Galilei, Galileo—An Italian scientist (1564–1642). He was the first scientist to introduce systematic quantitative measurements into scientific research, and thus formed the basis for experimental science and for the experimental testing of theories. He also was the first to introduce time as a parameter in physical processes, and by precise measurements of time intervals introduced systematic measurements of entities connected with motion, such as velocity and acceleration.

Among his important discoveries are the use of lenses to construct a telescope, by which he discovered that the moon's surface is not smooth, and he was the first to observe some of the moons of Jupiter. The statement that all freely falling objects fall with the same acceleration is related to Galilei.

Galilean relativity—The use of the principle of relativity, on the assumption that time is an absolute quantity and is independent of the material world or of the relative state of the observer, is the basis for Galilean relativity. Galilean relativity, together with Newton's assumption that space is an absolute entity independent of the material world, were the basis for Newtonian mechanics, which was commonly accepted by scientists for hundreds of years. Measurements connected with the velocity of light, that were performed toward the end of the 19th century (Michelson–Morley experiment), revealed contradictions in the law of addition of velocities which emerged from Galilean relativity. The solution to these contradictions was found in the Special Theory of Relativity, proposed by Einstein in 1905. The solution was based on the statement that time is not an absolute entity, and that the results of a measurement depend on the state of the observer relative to the measured phenomenon. The General Theory of Relativity showed that the measurements of time also depend on the distribution of matter in the universe.

General gravitation—The general gravitation includes the static gravitation, which is the gravitational attraction between masses as calculated by Newton, and the inertial phenomena, which are implied from the inertial property of matter and which appear when masses are subject to acceleration. From these phenomena, the centrifugal force and the Coriolis force emerge. The forces emerging from general gravitation, both the static force and those connected with acceleration, are proportional to the masses of the objects that are subject to these forces.

Geodesic—(Verbally—fitted to the earth's surface.) In differential geometry, the geodesic is the shortest line that connects two points on a surface, contained in the surface. In the General Theory of Relativity, the physical meaning of a geodesic is that a free-falling object moves along a geodesic in the four-dimensional space.

Homogenous—A situation in which no difference exists between different points of the relevant region.

Inertia—A property of material objects expressed in their "resistance" to changes in their state of motion (a rest or a motion with constant velocity). The amount of inertia of an object is proportional to its mass.

Infinitesimal—A small quantity chosen to be small up to any limit defined arbitrarily. E.g., when we calculate the length of a curved line, we divide the line into a very large number of small segments (infinitesimal segments), and thus we can treat each segment as a straight line. The length of the entire line is the sum of the lengths of all the infinitesimal segments. The summation of the segments' lengths is done by integration.

Integration—The limiting process of summation, where the summation is performed on a very large number of infinitesimal quantities. The expression in the integral is named "the integrand." The integration is the inverse operation of the differentiation. Hence, the integration sometimes can be performed by finding the primitive function of the integrand—the function whose derivative is the integrand. Since constant entities do not contribute to the result of differentiation, it should be remembered that while performing an integration, the solution may contain constant quantities in addition to the primitive function. The values of these constant quantities are determined by the boundary conditions—the values of the function at known conditions.

Interaction—Mutual action between two objects. Physically, any action is an interaction. When object a acts on object b, the object b acts on object a. Hence, in calculating the actions of an object, we should also include the backward action on the acting object.

Interval—The difference between the values of a certain quantity measured at two different points.

Space interval—The difference between the spatial coordinates of two events.

Time interval—The difference between the time coordinates of two events.

In the theory of relativity the concept of the interval has a specific meaning in a four-dimensional space. In the Special Theory of Relativity, the square of the interval (dS^2) is given by the difference between the square of the time interval (multiplied by c^2), and the square of the spatial interval. In the General Theory of Relativity, the square of the interval is given by:

$$dS^2 = A_{ij}dx_i \, dx_j,$$

where A_{ij} are the components of the metric tensor, and dx_i is the differential of the coordinate x_i. We distinguish between two types of intervals:
Time-like interval, where $dS^2 > 0$;
Space-like interval, where $dS^2 < 0$.

Invariant—An entity that remains unchanged (conserved) when its environment changes. An entity conserved on transforming from one coordinate system to another.

Isotropic—A situation in which no difference exists between observations made in different directions.

Laplacian—Designated by ∇^2. The expression ∇ designates a differentiation operator with respect to the spatial coordinates. ∇^2 shows that this operator operates twice consecutively on the same function. This operation results in second derivatives with respect to the spatial coordinates.

Linear approximation—A treatment of complicated equations in which terms containing products (or powers) of physical variables are dropped. This method suits the treatment of equations that contain variables whose values are very small. In such a case, the products (or the powers) of the small variables become much smaller, and they can be ignored, relative to the other variables, which are larger than the products by a few orders of magnitude. The equations then become linear, and their treatment is more convenient. It should be remembered that through such a treatment, the limits in which this treatment is valid should be kept.

Lorentzian—After Lorentz (1853–1928), who was the first to calculate the transformation equations that suited the Special Theory of Relativity. A Lorentzian frame of reference is an inertial four-dimensional frame, which can be represented by a four-dimensional flat coordinate system. Such a system is the appropriate system for the description of nature in accordance with the Special Theory of Relativity.

Lorentz operator—The operator that generates the Lorentz transformations of four-dimensional vectors. It transforms four-dimensional vectors from one inertial system to another. The action of this operator depends upon the relative velocity between the two systems. The matrix that represents this operator in the case in which the relative velocity v is along the x axis is given by:

$$
\begin{pmatrix}
\gamma & 0 & 0 & \frac{-\gamma v}{c} \\
0 & 1 & 0 & 0 \\
0 & 0 & 1 & 0 \\
\frac{-\gamma v}{c} & 0 & 0 & \gamma
\end{pmatrix},
$$

where γ is the Lorentz factor

$$
\gamma = \frac{1}{\sqrt{1 - \frac{v^2}{c^2}}}.
$$

This factor appears in the formulae used to calculate the contraction of length, the time dilation, and the increase of mass with motion in the Special Theory of Relativity.

Mach—An Austrian scientist who lived at the end of the 19th century and the beginning of the 20th century. The velocity of sound in air is called "Mach" after him.

Mach principle—This principle states that the inertial properties of masses are due to an interaction between the accelerated mass and all the other masses in the universe. The Mach principle was used by Einstein as a basis for the formulation of the Principle of Equivalence. To date, it is not commonly accepted in the scientific community that the Mach principle is valid.

Main diagonal—The main diagonal in a square matrix is the diagonal running from the upper left corner of the matrix to its lower right corner. The components of matrix **A** along the main diagonal are designated by A_{ii}. The symmetric properties of the matrix are defined with respect to the main diagonal.

Matrix—A two-dimensional mathematical form that contains components arranged in rows and columns. In a square matrix the number of columns equals the number of rows. A symmetric matrix is one in which each component equals its symmetric component with respect to the main diagonal. If we designate a matrix by **A**, the symmetry is manifested by the fact that $A_{ij} = A_{ji}$. An anti-symmetric matrix is one in which each component has the negative value of its symmetric component with respect to the main diagonal; i.e., $A_{ij} = -A_{ji}$. In such a matrix all the components contained in the main diagonal are zeros.

Momentum—The amount of motion. The momentum of an object equals the product of its mass times its velocity. The momentum characterizes the motion quantitatively. The original formulation of the second law of Newton is: The rate of change of the momentum of an object is equal to the sum of forces acting on this object. In a closed system, where no external forces act, the sum of the momenta of all the components of the system is a conserved quantity (the law of conservation of momentum). Certain parts of the system can interchange momentum through internal interactions, such as collisions, but these interchanges of momentum will always take place according to the law of conservation of momentum.

Angular momentum—A quantity that characterizes an angular motion. The angular momentum of an object is defined with respect to a certain axis of rotation. The angular momentum is equal to the product of the momentum of the object times its perpendicular distance from the axis of rotation. In a closed system, the sum of all the angular momenta of the components of the system is a conserved quantity.

Neutron—It is one of the basic constituents of the atom. A neutral particle whose mass approximately equals the mass of the proton.

An isolated neutron is unstable. Its life time is about 1000 seconds, and it disintegrates to a proton, an electron, and a neutrino. The neutrons located inside the nucleus of an atom are stable.

Newton, Isaac—An English scientist (1642–1727). Known especially for the laws of mechanics and the law of gravitation, which he discovered. From these laws he calculated the equations for the orbits of the planets. He discovered many other laws in diverse fields of physics.

The law of gravitation is the law that defines the gravitational attraction between any two masses. The magnitude of this force is proportional to the product of the two masses and inversely proportional to the square of the distance between them. This interaction between the masses is the *static* gravitational interaction.

Newton's laws of motion:—

The first law—Also referred to as the law of inertia. It states that when the sum of the forces acting on a certain object vanishes, the object will sustain its state of motion—either at rest or at a motion with a constant velocity in a straight line.

The second law—Also referred to as the law of motion. It states that the rate of change of the momentum of an object equals the sum of forces acting on this object. In another formulation, which emphasizes the role of acceleration, this law is formulated as: the acceleration of an object is proportional to the sum of the forces acting on this object and inversely proportional to the mass of the object.

The third law—Also referred to as the law of action and reaction. It states that in an interaction between two objects, a and b, the force that a exerts on b is equal in magnitude and inverse in direction to the force that b exerts on a.

Osculatory—An osculatory circle to a line at a given point is the circle that coincides with the line at that point. An osculatory plane is the plane that contains the osculatory circle.

Parameter—A measure, quantitative characterization. A quantity used as a measure for a series of quantities. A quantity that characterizes quantitatively a group of entities that have something in common.

Perihelion—In an orbit of a planet around the sun, the point at which the planet is closest to the sun.

Principle of complementarity—This principle was stated by Niels Bohr for the connection between quantum and classical mechanics. It states that results calculated according to quantum mechanics will correspond to the results calculated by classical mechanics when dealing with a large number of particles or with average results over long time intervals.

   In parallel, a complementary principle can be stated for the theory of relativity. The results calculated according to the General Theory of Relativity will correspond to the Newtonian gravity when dealing with weak gravitational fields and low velocities.

Proper length—A proper length of an object is its length as measured in the rest system of the object.

Proper time—The proper time of an object is the time measured by a clock located in the rest system of reference of the object.

Pulsars—Celestial objects that radiate pulses at high frequency in the range of radio wavelength. They were discovered during the 1960s. We know now that pulsars are neutron stars rotating with very high angular velocity, and the frequency of their pulses is the frequency of their rotation. Some of the pulsars were found in or close to the remnants of supernova explosions, and it is believed that they were formed during the explosion.

Quadrupole—Four-poles object. A system whose symmetry contains four poles.

Quarks—It is believed that the elementary particles we know (protons, neutrons, etc.) are built from subparticles called quarks. (The name quark has no meaning.) The electric charge of a quark is a fraction (1/3, 2/3) of the elementary charge of the electron. By defining the laws of interaction between different quarks the entire system of the elementary particles can be constructed. To date, no way has been found to isolate a single quark.

Quasars—Quasi-stellar radio sources. Celestial objects discovered during the 1960s. The large red shift found in their radiation ($Z > 1$) shows that they are located at very large distances. The amount of energy that they radiate is a few orders of magnitude larger than the energy radiated by an entire galaxy. There is no explanation yet for the source of this amount of energy.

Radius of curvature—The radius of curvature of a line at a given point is the radius of the circle that is tangent (osculates) to the line at that point, and the segment of the line at that point is a part of the oscullating circle.

The radius of curvature of a surface at a given point is calculated from the radii of curvature of lines contained in the surface and intersecting at the point.

Rank of a tensor—Defines the number of *dimensions of the form* that represents the tensor. The number of the components of a tensor equals the number of dimensions of the space, raised to the power of the rank of the tensor. Let us designate by n the number of dimensions of space. A scalar is a tensor of zero rank. Hence the number of its components is $n^0 = 1$. A vector is a tensor of rank 1. Hence the number of its components is $n^1 = n$. A matrix represents a tensor of rank 2, and the number of its components is n^2. Generally, there exist tensors with any value for their rank.

Refraction index—The refraction index of a medium is a number that expresses the ratio of the velocity of light in vacuum to the velocity of light in the medium. When a light passes from a medium with a certain refraction index to a medium with another refraction index, it experiences a change in its velocity. As a result, its trajectory is bent according to Fermat's Principle. This principle states that the trajectory of light on its passage from one point to another is such that the time needed for this passage is minimal.

Shift of the spectrum—A change of the wavelength (and the frequency) of a wave phenomenon. We know the Doppler shift of the spectrum, created by the relative velocity between the receiver and the emitter. As well, we know the gravitational shift of the wavelength of light, created by gravity, when there is a potential difference between the locations of the emitter and the receiver.

Specific angular momentum—Angular momentum per unit mass.

Specific energy—Energy per unit mass. The specific energy of an object is obtained by dividing the total energy of the object by its mass.

Spectral distribution—A characteristic spectrum of a luminous object. The spectral distribution of an object indicates the radiation intensity at each wavelength of the radiation. The spectral distribution is determined by the temperature and the velocities of the particles at the radiating surface, and by the chemical composition.

The spectral distribution is specific for each object and is used for "personal" identification of luminous objects.

Subspace—In a space of a given number of dimensions, we can always define a subspace by assuming a constant value for one (or more) of the coordinates. The domain defined in this way is a space whose number of dimensions is lower by one (or more) than the number of dimensions of the original space, and it is a subspace of the original space.

Tidal forces—The forces acting on an object, created by the difference between the gravitational forces that act at different points in the object.

Transformation—Change of form. In our concern—a transition from one coordinate system to another. The transformation equations are the equations that describe how we transfer information (results of observations) from one system of reference to another. The Lorentz transformations are the transformations between inertial systems.

From the transformation equations between a local flat system to the system of the "real" space at the same location, we can calculate the components of the metric tensor of the "real" space at that location.

Variations—Small changes. The method of variations is a method used for finding extremal trajectories. According to this method, we vary the integration path by small variations, keeping the edge points of the path constant. When we find a path that small variations do not change the value of the integral along the path, this is the extremal path.

Vector—A mathematical entity that has both a magnitude and a direction. It is designated by an arrow, whose length is proportional to the magnitude of the vector, and the direction it points is the direction of the vector. We use geometrical vectors to represent physical quantities that have a magnitude and a direction (such as force, velocity, etc.). In a given coordinate system, the vector is given by its components along the coordinate axes. The number of the components of a vector equals the number of dimensions of the space.

Scallar product of vectors—A product of two vectors whose result is a scalar. In a Euclidean space, the scalar product is achieved by multiplying each component of one vector by the corresponding component of the second vector and then finding the sum of all these products. In non-Euclidean systems, the scalar product is achieved with the aid of the metric tensor. (In a Euclidean system the metric tensor is represented by the unit matrix.)

Vectorial sum of vectors—Finding the vector whose action is equivalent to the actions of all vectors included in the summation. When the summed vectors do not act exactly in the same direction, the length of the equivalent vector is smaller than the algebraic sum of the lengths of all the vectors. Each component of the equivalent vector found by the vectorial addition equals the sum of the corresponding components of the summed vectors.

World vector—A four-dimensional vector as defined in the theory of relativity. It contains three spatial components (which are analogous to the common three-dimensional vectors), and a fourth dimension component, parallel to the time axis. In the Special Theory of Relativity, the world vectors are transformed from one inertial system to another by the Lorentz operator. The scalar product of world vectors is performed by first multiplying one of the vectors by the metric tensor, and then multiplying the vector obtained from this product by the second vector. A scalar product yields a scalar that is invariant under transformations between coordinate systems.

White dwarf—A star that has completed the evolutionary phase of nuclear burning. It cools down, and since the thermal pressure decreases in its interior, it cannot support the star against the gravitational force. The star contracts and reaches a very high density at its center—above 10^5 gram cm^{-3}. At such a high density, the distances between the atoms are of the order of the radius of a regular atom. The atoms have lost most of their electrons, and these electrons are free to move inside the matter. The pressure created by the free electrons emerges from a phenomenon called "degeneracy," and it does not depend on the temperature. This degeneracy pressure can support the star against gravity if the stellar mass is below $1.4 M_\odot$. (This mass limit is called the "Chandrasekhar mass.") A cool star, whose mass is above this limit, will collapse further and will form a neutron star.

World line—The trajectory of an object as drawn in a space–time diagram. A world line of an object is the list of all the events in which this object was involved during its history.

Index

accretion disc, 196
addition of velocities, 8
apparent, 44, 45, 48

big bang, 186, 204
black hole, 156, 164, 193, 194
Bohr, 120

Cartesian, 25, 26, 28, 31
causal, 64
centrifugal potential, 122, 160
Chandrasekhar, 194
contravariant, 38
Coriolis, 88, 125
covariant, 39, 70, 77, 119
Coxeter, H. S. M., 54
Cygnus, 195

D'Alembert, 96
Descartes, 25
differential, 32, 34, 36, 54, 71
dipole, 200
disc, 71–73, 122, 125, 126, 128
Do Carmo, 54
Doppler, 154, 182

Eddington, 151
Einstein, 12
Einstein's ring, 153
elevator, 11, 87, 88, 121
equation of state, 183, 194
equivalence, 11, 74, 86, 88, 102, 155
ether, 9

feedback, 171
Fermat, 113, 152
Feynman, 116
Friedman, 172

Galilean, 5
Gaussian, 47, 48, 175, 176
geodesic, 50, 109, 111, 116, 117, 123, 125,
    128, 143, 145, 161, 179
gravitational lense, 152, 192

Hubble, 182, 185, 190

infinitesimal, 32, 119
interaction, 78, 80, 81, 95, 96, 129, 169, 170,
    204
interval, 60, 63, 72

invariance, 24
invariant, 24, 51, 60, 66, 123
isotropic, 42
isotropy, 172, 175

Kreyszig, 55

Laplacian, 131
Leibniz, 23
Leverrier, 148
Lorentz, 63, 69, 197

Mach, 95, 99
Maxwell, 132
Michelson, 9
Minkowski, 66
Misner, 202, 204
Moller, 75

neutron stars, 159
normal, 20, 47
Novikov, 204

osculating, 46, 111

Poisson, 130
polar, 31, 33, 34
Pound, 94, 154
Poynting, 83
proper length, 64
proper time, 64
pseudo-rotation, 62, 101
pulsar, 195, 201
Pythagorean, 27, 28

quadrupole, 200, 202
quasar, 191
quasars, 190

radius of curvature, 46, 177
rank, 21
Rebka, 94, 154
Riemann, 66
Rindler, 71, 75

saddle, 48
scalar product, 13, 15, 26, 28, 60, 68
Schwarzschild, 137, 139, 141, 146, 156, 157,
    193, 196
Sciama, 100
Snell, 115
spacelike, 64, 158
stress tensor, 20, 131
Szamosi, 8, 12

Thorne, 202
timelike, 63, 158
Tolman, 95
transformation, 7, 24, 62, 69, 70, 101

Weber, 201, 202
Weinberg, 204
Wheeler, 202, 204
white dwarf, 94, 194
world line, 58
Wrede, 54, 55

Zeldovitch, 204